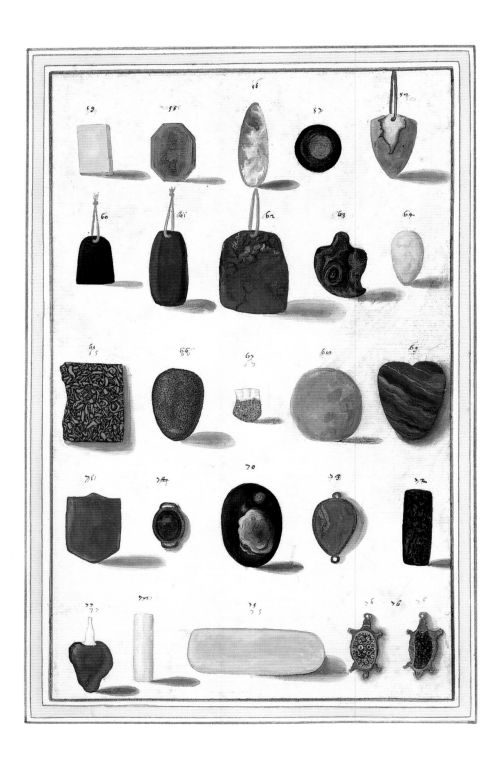

AMAZING

RE THINGS

THE ART OF NATURAL HISTORY
IN THE AGE OF DISCOVERY

DAVID ATTENBOROUGH

SUSAN OWENS, MARTIN CLAYTON
AND REA ALEXANDRATOS

YALE UNIVERSITY PRESS
NEW HAVEN AND LONDON

Published 2007 in the United States by Yale University Press and
in the United Kingdom by Royal Collection Enterprises Ltd.

Designed by Mick Keates
Production by Debbie Wayment
Project management by Paul Holberton

Printed in Italy by Studio Fasoli, Verona
Printed on Symbol Tatami Ivory, Fedrigoni Cartiere Spa, Verona

Library of Congress Control Number: 2006938185
ISBN 978-0-300-12547-4 (cloth : alk. paper)

A catalogue record for this book is available from the British Library.

The paper in this book meets the guidelines for permanence and durability of the Committee
on Production Guidelines for Book Longevity of the Council on Library Resources.

10 9 8 7 6 5 4 3 2 1

Front cover: Maria Sibylla Merian, Passion flower plant (plate 59, detail)
Frontispiece: from the Paper Museum of Cassiano dal Pozzo, Gems, stones and amulets (plate 33)

CONTENTS

FOREWORD

In the early years of the sixteenth century Leonardo da Vinci observed that 'The eye is the chief means whereby the understanding may most fully and abundantly appreciate the infinite works of nature'. Using his eyes and his intelligence, an artist could accurately and concisely record reality, whereas to do so in words alone would take 'a confused prolixity of writing and time'.

The use of imagery to document the astonishing diversity of nature has passed through many stages in the past five hundred years. Recently the possibilities opened up by new techniques of photography and filming have seemed endless. Some of the results have been brought to the general public by the naturalist and broadcaster Sir David Attenborough, one of the pioneers of the nature documentary (and the man responsible for the introduction of colour television into Britain in 1965). This publication is the result of a most fruitful collaboration between the Royal Collection and Sir David, who assisted the curators of the Collection in the choice of works for the exhibition, and has contributed the introductory essay and detailed comments on many of the works.

The eighty-seven watercolours considered here come from five exceptional groups of natural history drawings and watercolours in the Royal Collection. They date from a 250-year period between the late fifteenth and early eighteenth century – the Age of Discovery, when European knowledge of the world was transformed by voyages to Africa, Asia and the Americas. Leonardo da Vinci (1452-1519) was working at the dawn of this era, recording the familiar flora and fauna of Italy but with the new spirit of scientific investigation that marked the Renaissance. Over a hundred years later, Cassiano dal Pozzo (1588-1657) in Rome and Alexander Marshal (c.1625-1682) in London catalogued many new species that were being brought by explorers and traders to the shores of Europe. And the artists Maria Sybilla Merian (1647-1717) and Mark Catesby (1679-1749) travelled to the New World themselves, to record exotic and previously unknown creatures in their natural habitats.

As a glance through the pages of this book will show, the ways in which these artists recorded nature are very diverse. Leonardo's sensitivity and incisiveness of line is a world away from what Catesby described as his own 'Flat, tho' exact manner'. The exuberant curves and spirals with which Merian endowed the 'amazing rare things' that she found in South America are a stylistic trait which make her work unmistakable. The shared ground, however, is the artists' extraordinary engagement with the natural world, whether they made pioneering expeditions to the Americas or found their subjects in their own garden. Through painstaking examination and description, all these artists hoped to comprehend the riches of the natural world.

The five groups of drawings entered the Royal Collection at different times. The 600 sheets by Leonardo were acquired in the late seventeenth century, almost certainly by Charles II, the founder of the Royal Society. Three of the groups of drawings – 95 by Merian, 2,500 from the Dal Pozzo collection, and over 250 by Catesby – were acquired by George III in the mid-eighteenth century as part of his library collection. And Marshal's florilegium, containing over 150 watercolours, was presented to the future King George IV early in the nineteenth century.

The Leonardo drawings have been accessible to scholars for many years, particularly through the catalogues by Kenneth Clark and Carlo Pedretti (1968-9). Marshal's florilegium was the subject of a magnificent volume by Prudence Leith-Ross published by the Royal Collection in 2000, and the contents of Cassiano's 'Paper Museum' are being published in a series of catalogues which began to appear in 1996. It is intended that the drawings by both Merian and Catesby will in due course be published in full. The essays introducing the five groups here are the work of curators in the Print Room at Windsor: Martin Clayton (Deputy Curator) has written on Leonardo da Vinci, Rea Alexandratos (Dal Pozzo Project Coordinator) on Cassiano, and Susan Owens (Assistant Curator) has written the essays concerning Marshal, Merian and Catesby, and has skilfully guided the publication through its production stages.

I would like to express our sincere gratitude to Sir David Attenborough for his extensive contribution to this project, and for so generously sharing with us his remarkable knowledge and insight into the natural world. His involvement from its very beginning has been truly invaluable.

We should also like to thank the following for their generous help: Giulia Bartrum and Kim Sloan at the British Museum; Julie Harvey at the Natural History Museum; Kay Etheridge at Gettysburg College; Tessa Rankin, Ella Reitsma and Esther Schulte. We are indebted to Prudence Leith-Ross for her comprehensive study of Marshal. We also gratefully acknowledge the fundamental work on Cassiano, Catesby and Merian carried out by Henrietta McBurney, formerly Deputy Curator of the Print Room.

JANE ROBERTS
Librarian and Curator of the Print Room

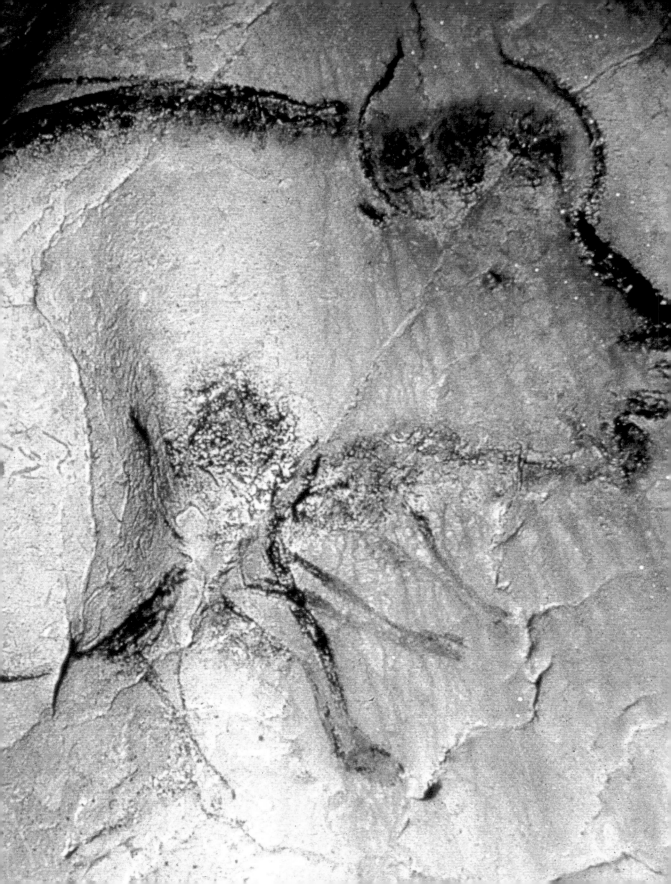

Picturing the
Natural World

David Attenborough

Animals were the first things that human beings drew. Not plants.
Not landscapes. Not even themselves. But animals. Why?

The earliest of all known drawings are some thirty thousand years old. They survive in the depths of caves in western Europe. The fact that some people crawled for half a mile or more along underground passages through the blackness is evidence enough that the production of such pictures was an act of great importance to these artists. But what was their purpose? Maybe the act of drawing was an essential part of the ceremonials they believed were necessary to ensure success in hunting, for some of the creatures represented appear to be wounded or disembowelled and others have chevrons scratched across them that could be interpreted as spears. Maybe the paintings were intended not to bring about the death of the creatures portrayed but, on the contrary, to ensure their continued fertility so that the people would have a permanent source of meat. We cannot tell.

One thing, however, is certain. These drawings are amazingly assured, wonderfully accurate and often breathtakingly beautiful. Those who produced them had observed

Fig. 1
*A bison, c.*30,000 BC
Cave of Chauvet-Pont-d'Arc, France

their subjects with such intensity that they could draw the images, in the flickering light of their torches, entirely from memory. The grace of a galloping horse, the swell of a gravid mare's belly, the overwhelming strength and power of a charging bull, all are portrayed with supreme skill (fig. 1).

This practice of painting images of animals on walls has persisted throughout mankind's history. Five thousand years ago, when men in Egypt began to build the world's first cities, they too inscribed images of animals on their walls. There is no doubt about the function of at least some of these. The Egyptians worshipped animals. Sacred bulls were mummified and enclosed in immense granite sarcophagi. Ibis and falcons were sealed in terracotta pots and stacked in long galleries by the tens of thousands. And when images were painted on temple walls, their god-like status was made plain by giving them partly human features. So sometimes hawks, apes and crocodiles, while they have fully recognisable animal heads, have been given human torsos. But the

Fig. 2
A fragment of a wall painting from the tomb of Nebamun, c.1350 BC From Thebes, Egypt

Fig. 3
*A fragment of a wall
painting from the tomb
of Nebamun, c.1350 BC
From Thebes, Egypt*

Egyptian artists also delighted in the straightforward natural beauty of animals, for they adorned the walls of their own underground tombs with pictures of red-breasted geese rising from the papyrus swamps, of cats stalking birds, even butterflies fluttering between palm trees (fig. 2). The mummified dead in the next world would surely wish to be reminded of the beauties and delights of this one.

Plants appear in these paintings. There are none in the ancient caves. Those people who drew rampaging bulls and fleeing deer had not yet discovered how to sow seeds and reap the resulting crops. The Egyptians, however, were farmers and they valued plants. They knew how to discriminate between different kinds and they displayed that knowledge on the walls and in their manuscripts. They drew cultivated gardens, vines laden with grapes and fields of wheat rich with grain (fig. 3).

The distinction between animals as gods and animals as themselves, so vivid and clear in ancient Egypt, is also apparent in the manuscripts of early Christianity. The medieval monks, who five thousand years after the fall of Egypt sat in the scriptoria of their monasteries, embellishing the capital letters of their manuscripts with intricate interlacings, also provided their saints with emblematic animals. Saint Mark has his lion, but it is a lion with wings; and Saint John is accompanied by an eagle but often one of such magnificence that it is scarcely recognisable as the brown bird of reality.

Nonetheless, other less exalted and more earthy creatures crept into their manuscripts. The scribes, perhaps as an escape from the solemnity of their devotions, indulged their imaginations and affections by introducing on to their pages the wild creatures that abounded in the natural world outside. Squirrels run up the margins and rabbits chase one another around the limbs of the capital letters. The artists also indulged their sense of humour. A giant snail, tentacles outstretched, jousts with a knight in armour (fig. 4); a dog pumps a chamber organ which is being played by a rabbit; an ape posing as a doctor offers medicine to a prostrate and presumably sick bear (fig. 5).

Fig. 4
A knight and a snail duelling
From the Macclesfield Psalter, *c.*1330

In the early twelfth century, such animals began to escape from the breviaries and Psalters into books of their own. These bestiaries seem to be a particularly English phenomenon. Of the sixty-five examples known, fifty originated in England (fig. 6), an early indication perhaps of a special national affection for animals that endures to this day. But the animals have not yet escaped from their religious connections. The bestiary texts explained that animals were put on earth to illustrate God's purpose and teachings. They were, in the eyes of the devout, parables and sermons, and their morality was even more important than their morphology. Thus the reason a wolf's eyes shine in the dark is to demonstrate that many things that seem attractive are in fact the works of the devil. Among these recognisable images there are some fantastic animals – unicorns and dragons, sea monsters, and griffins that were part-lion and part-eagle (fig. 7). The scribes had not seen them but they certainly believed in their existence.

In the fifteenth century, however, the new scientific spirit of the Renaissance swept through Europe. Scholars began to examine the world with fresh eyes and to question the myths and fantasies of the medieval mind. Galileo scanned the skies with the newly invented telescope to ponder the movements of the planets. And Leonardo da Vinci, the

Fig. 5
An ape doctor with a bear patient
From the Macclesfield Psalter, *c*.1330

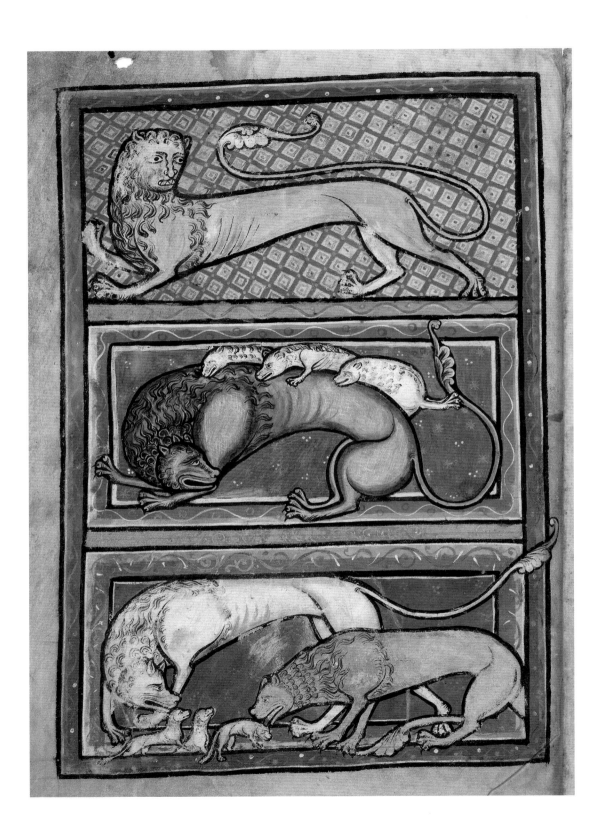

Fig. 6
Lions with their cubs
From an English bestiary, *c.*1200

most dazzling Renaissance man of all, started to look at the bodies of animals and plants in a new way. He wished to understand how they grew, moved and reproduced themselves, so he not only drew them in action but dissected their bodies. As a professional artist, he needed to include horses in his paintings and sculptures. To do that he needed to know how their muscles powered their skeletons. Only if he understood that could he adequately delineate the swellings and grooves they created on the surface of a horse's body (plates 4, 5 and 18). Bears were not such a common professional requirement for his compositions but they nonetheless excited his curiosity (fig. 8). How exactly did they walk? He dissected the foot of one, isolated its tendons, and then drew what he had exposed to help him work out the mechanics involved (plate 6). He watched a cat stalking a bird and recorded its movements with meticulous accuracy.

Fig. 7
A griffin
From an English bestiary,
*c.*1230–40

He also drew dragons. Did he believe that they really existed? Whether he did or not, his understanding of the mechanics of animal movements enabled him to sketch on the same sheet of paper that he used to draw cats as convincing a dragon as you can find (fig. 9).

He also turned his enquiring mind to plants. Here his concern is sometimes with their detailed structure; at other times he is more interested in their characteristic patterns of growth. His sketches of blackberries, for example, describe every element of the individual fruits (plates 9 and 12). And yet his study of the Star of Bethlehem (plate 7) is distinctly stylised, emphasising the curling growth of the new leaves.

Fig. 8
Leonardo da Vinci
*A walking bear, c.*1490

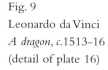

Fig. 9
Leonardo da Vinci
*A dragon, c.*1513–16
(detail of plate 16)

Other scholars began to assess the bewildering variety of animal life that lay beyond their own countryside. People had known from Roman times that away to the south, across the Mediterranean in Africa, there were huge antelopes with necks twice the length of their legs, and horses that were patterned with black and white stripes. And there were stories, that to their minds were doubtless equally credible, of mermaids who rose from the waves with babes at their breast, of seven-headed monsters that breathed fire, of tribes of human beings with no heads but mouths in the middle of their stomachs and every other conceivable anatomical variation (fig. 10). But now explorers travelling south down the coast of Africa, east to the Indies and west to the New World were bringing back completely new kinds of creatures to dazzle and baffle European naturalists. The natural world that was full of so many newly discovered wonders needed a catalogue.

The first to compile one, and thereby earn himself the title of the father of zoology, was a Swiss doctor, Conrad Gesner. To illustrate it, he assembled drawings from wherever he could get them. Some he commissioned from artists, who drew many of the skins, skeletons and other objects that Gesner himself had collected in addition to some familiar animals (fig. 11). Others he asked to use their imaginations to produce

Secunda etas mundi · Folium · XII

De homib[us] diuersar[um] forma[rum] dicit Pli. li. vij. ca. ij. Et Aug. li. xvi. de ci. dei. ca. viij. Et Isidorus Ethi. li. xi. ca. iij. oia q̃ sequitur in india. Cenocephali homines sunt canina capita habentes cũ latratu loquũtur aucupio viuũt. vt dicit Pli. qui omnes vescũtur pellibus animalũ.

Ciclopes in India vnũ oculum bñt in fronte sup nasum hij solas ferarũ carnes comedũt. Ideo agriofagite vocãtur supra nasomonas confinesq̃ illozũ homines esse:vtriusq̃ nature inter se vicibus coeũtes. Calliphanes tradit Aristotiles adijcit dextram mamam ijs virilem leuam muliebrem esse quo hermofroditas appellamus.

Ferunt certi ab orietis pte intima esse homines sine naribus:facie plana eq̃ li totius corpis planicie. Alij os supiore labro orbas. alios sine linguis ⁊ alijs cõcreta ora esse modico foramine calami auenar[um] potũ hauriẽtes.

Item homines habentes labiũ inferius.ita magnũ vt totam faciem contegant labio dormientes.

Item alij sine linguis nutu loq̃ntes siue motu vt monachi.

Pannothi in scithia aures tam magnas bñt. vt contegant totum corpus.

Artabrite in ethiopia pni ambulãt vt pecora. ⁊ aliqui viuũt p annos.xl. quẽ nullus supgreditur.

Satiri homũciones sunt aduncis naribus cornua i frontibus bñt ⁊ capraz pedibus similes qualẽ in solitudine sanctus Antonius abbas vidit.

In ethiopia occidentali sunt vnipedes vno pede latissimo tam veloces vt bestias insequantur.

In Scithia Ipopedes sunt humanã formaz eqnos pedes habentes.

In affrica quasdã familias effascinantũ Isigonus ⁊ Memphodorus tradit quaz laudatõe intereãt, pbata.arescãt arbores: emoriãtur infantes. esse eiusdem generis in tribalis et illirijs adijcit Isogon[um] q̃ visu quoq̃ effascinẽt iratis pcipue oculis: quod eo rũ malũ facilius sentire puberes notabili[us] esse cp pupillas binas in oculis singulis habeant.

Item boies.v.cubitor nũcp infirmi vsq̃ ad mortes. Hec oia scribũt Pli. Aug. Isi. Preterea legit i gestis Alexãdri cp i india sunt aliq̃ boies sep man[us] bñtes. Itẽ boies nudi ⁊ pilosi in flumine morãtes.

Itẽ boies manib[us] ⁊ pedib[us] sex digitos habentes. Itẽ apothami i aq̃s morãtes medij boies ⁊ medij caballi.

Item mulieres cũ barbis vsq̃ ad pect[us] § capite plano sine crinibus.

In ethiopia occidẽtali sũt ethiopes.iiij.oclos bñtes. In Eripia sunt boies formosi ⁊ collo gruino cũ rostris aialium boimq̃ effigies mõstruferas circa extremitates gigni mime mirũ. Artifici ad formanda corpora effigiesq̃ celandas mobilitate ignea.

Antipodes,at ee.i.boies a contraria pte terre:vbi sol oritur qñ occidit nob aduersa pedib[us] nris calcare vestigia nulla rõe credẽdũ ẽ vt ait Aug.16.de ci. dei.c.9. Ingẽs tñ b pugñ lřaz cõtraq̃ vulgi opiões circũfundi terre boies vndicq̃ couersiliq̃ iter se pedib[us] stare et cũctis similem eẽ celi vtice. Ac sili mõ et q̃cũq̃ pte mediã calcari. Cur at ñ decidãt:mireĩ ⁊ illi nos ñ decidere: nã eĩ repugnãte:⁊ quo cadãt negãteyt possint cadere. Nã sic ignis sedes nõ ẽ nisi i ignib[us]:aq̃rũ nisi i aq̃s. spũs nisi in spũ. Ita terre arcentibus cũctis nisi in se locus non est.

pictures of animals that neither they nor anyone else had ever seen (fig. 12). And some images he simply appropriated. Among these was a picture of an Asiatic rhinoceros. The animal itself had arrived in Lisbon in 1515, a present to the King of Portugal from Goa, his recently conquered territory in India. A drawing of this sensational creature came into the hands of the great artist Albrecht Dürer, who lived in Nuremberg, which since medieval times had been a centre for the manufacture of armour. He redrew the sketch and in the process gave the animal spectacular armour-plating, complete with bosses and cuirass and a small supernumerary horn on its shoulders (fig. 13).

All these pictures, together with long verbal descriptions and quotations, Gesner published from 1551 onwards in the four hefty volumes of his *Historia animalium*. The entries were arranged not in groups based on their obvious relationships, but alphabetically. As a consequence, when his great work was translated into other languages, the animals had to be reordered – equally illogically. Thus when the English version of some of the volumes appeared, translated by Edward Topsell and entitled *A Historie of Four-footed Beastes*, the first entry, 'Antelope', was followed by 'Ape'.

Gesner's illustrations in turn were plundered by an Italian encyclopaedist, Ulisse Aldrovandi, the professor of natural sciences at the University of Bologna. Like Gesner, Aldrovandi amassed an immense collection of skins, skeletons, fossils, dried plants, insects and much else besides. His cabinet of curiosities was said to contain 4,554 drawers of specimens. For years he wrote and rewrote descriptions of his objects, quoting from all kinds of sources, including references he happened to discover in the verses of classical poets.

The myths of medieval times had not yet entirely vanished. Aldrovandi firmly believed in the existence of dragons and he accordingly devoted a section to them in which he enumerated their different species. Some had wings and some not. Some had seven heads. And he illustrated examples with eight legs, two legs (fig. 14) and none whatsoever, like snakes.

Nonetheless, there is a noticeable advance in Aldrovandi's scientific thought. He abandoned Gesner's alphabetical arrangement and instead adopted a more logical basis for his entries. He grouped all birds together in three volumes, though he included bats among them. He examined the skeleton and internal anatomy of a parrot, paying special attention to the organ that enabled the bird to speak, and he dissected the head of a

Fig. 11
A fox
From Conrad Gesner,
Historia animalium,
1551

Fig. 12
A unicorn
From Conrad Gesner,
Historia animalium,
1551

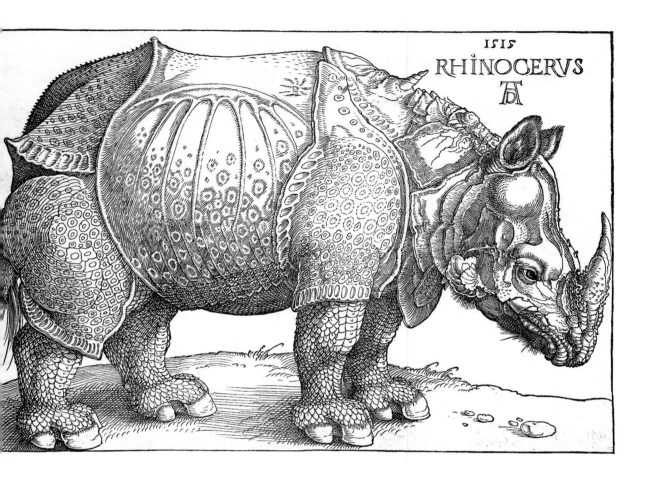

Fig. 13
Albrecht Dürer
A rhinoceros, 1515

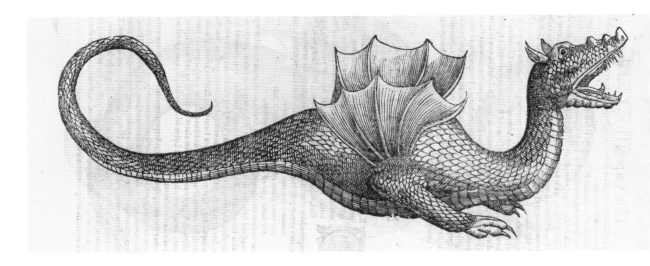

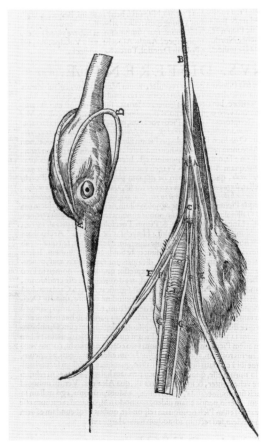

Fig. 14
A winged dragon
From Ulisse Aldrovandi, *Liber serpentium et draconum*,
1640

Fig. 15
The skull and tongue of a woodpecker
From Ulisse Aldrovandi, *Ornithologiae*, 1599

woodpecker and accurately showed how the socket for its immensely long tongue curled right round its skull (fig. 15).

His illustrations, like those in most printed books of the time, came from wooden blocks on which the design was cut in relief. The simplicity of the technique inevitably imposed limitations on detail. Many of Aldrovandi's images are banal representations of banal subjects – a tooth, indeterminate worms, or lumps of almost featureless stone. Some are simply inept. His zebra wears a puzzled, almost resentful look on its face as though it were baffled as to why anyone should paint its coat with such absurd stripes (fig. 16). But there are also many that have an undeniable grandeur. Snakes tie themselves into spectacular knots and huge crabs, shown full-page, glower at the reader from within their splendidly armoured and articulated shells.

Aldrovandi did not start publishing this great encyclopaedia until he was seventy-seven years old and he died after three volumes on birds and one on insects had been completed. His pupils, however, continued to edit the drawings and descriptions that he had left, and over the following years more and more volumes appeared. The last was printed in 1667, over sixty years after its author's death.

Fig. 16
A zebra
From Ulisse Aldrovandi,
De quadrupedibus
solidipedibus, 1616

Aldrovandi's works were firmly based on his own collections of specimens. In the first half of the seventeenth century, however, a new kind of collector appeared, who accumulated not objects but drawings of them. Such a one was Cassiano dal Pozzo. His interests were wide – classical statues and architecture, plants, animals, fossils. These drawings, which ultimately would fill numerous huge vellum-bound volumes, constituted what he called his *museo cartaceo*, his 'Paper Museum'. He was more rigorously scientific than Aldrovandi. Gone are the mermaids, dragons and centaurs. Cassiano's artists did not work from hearsay, but from physical actuality, although sometimes, faced with yet another curiosity from overseas, they had difficulty in making sense of what was in front of them. The sloth clearly baffled them. How could they be expected to imagine that in the great tropical forests of the New World there were animals that spent their lives hanging upside down from the branches of trees? That surely would be to revert to the fantasies of the medieval mind. So they drew the sloth gawkily trying to stand upright on its knuckles (plate 24).

Cassiano himself investigated the anatomy of many of his animal specimens as Leonardo had done a hundred and fifty years earlier – though the records his artists produced are no match for Leonardo's. He clearly believed, not unreasonably, that abnormalities could shed light on the true character of the normal. So he commissioned his artists to draw the misshapen and grotesque citrus fruits that he specially sought and valued (plates 20 and 32). His artists served him well and produced drawings which convey with great mastery the colour, texture and details of their subjects. Occasionally they succeeded in giving their drawings of the most ordinary objects a monumentality which the originals probably never possessed.

Gardeners throughout Europe shared this desire to catalogue their treasures – an understandable wish, bearing in mind the transitory nature of their plants and particularly their flowers. Such great catalogues of plants became known as florilegia. Most famously, a German archbishop, Johann Conrad von Gemmingen, commissioned an artist, Basilius Besler, to record all the plants in his garden near Eichstatt.

Fig. 17

A crown imperial

From Basilius Besler, *Hortus Eystettensis*, 1613

Corona Imperialis
Polyanthos.

Besler responded with a superb work, his *Hortus Eystettensis*, published in 1613, in which the plants were represented in splendid, almost heraldic positions that displayed their distinctive characteristics (fig. 17).

Besler's illustrations were reproduced not by woodcuts but by copperplate engravings, which could carry much finer detail. Many florilegia, however, remained in manuscript. Painting flowers, it seems, was in itself a kind of private worship of the glory of nature. For some painters, indeed, it became a pleasure so private that during their lives they kept the records of their devotions undisplayed to any but a few close friends.

One such was Alexander Marshal. He was a seventeenth-century Englishman of private means who devoted himself to horticulture and compiled his own florilegium entirely for his own delight. It was his way of relishing every detail of structure, every dapple and flush of colour. Occasionally, like the medieval monks, he also included alongside his tulips, irises and carnations illustrations of other subjects that intrigued and entertained him. Parrots and monkeys appear several times, for example, and like the monks' marginalia these extra creatures exist in a little world of their own, quite out of scale with the proper subject of the page (fig. 18).

Throughout the sixteenth and seventeenth centuries, the influx of strange and wonderful plants and animals from overseas supplied the artists of Europe with abundant new subjects. Eventually the time came when some decided that they would themselves travel to far places to find animals and plants in their original homes and settings. One of the first of these adventurous artists was, on the face of it, one of the least likely – a divorced middle-aged woman. Her name was Maria Sibylla Merian and she lived in Frankfurt, and later Amsterdam, where she earned her living as a flower painter, a teacher and a dealer in paints and pigments. She also had a passion for insects, and from an early age she had studied the life-cycles of moths and butterflies. In 1699, at what was then regarded as the advanced age of fifty-two, she set out with her daughter Dorothea Maria for the Dutch colony of Surinam in South America. She wanted to discover and chronicle for her own interest the stages through which insects passed during their life cycles. And she drew them, sitting on their particular food plants, often with other small and quite unrelated invertebrates alongside them.

Merian's work is unmistakable. Style in natural history drawing is often subtle – sometimes indeed it is almost unidentifiable. The artists are often too concerned with

Fig. 18
Alexander Marshal
A 'mexcico munky'
*(marmoset), c.*1650–82
(detail)

correctness and accuracy of detail in what they are portraying to mould its contours to their own particular taste. There is no room for vague impressions. There are no costumes to bring a sense of period. Since backgrounds are often omitted there is no need to use the conventions of perspective. So a drawing of a flower made in the sixteenth century may be hard to distinguish from one of the same species made in the twentieth. But Merian's pictures display clear clues to their authorship. These go beyond the characteristic mixture of flowers and insects. She also has a special fondness for curls and she draws them whenever she has the opportunity. The antennae of her moths all curl. So do the tendrils of her passion flowers, the tails of her lizards and the roots of her sweet potatoes. Given the chance, her snakes squirm into extravagant coils (fig. 19).

Fig. 19
Maria Sibylla Merian
A coral snake, *c.*1701–5 (detail)

Twenty years later and farther north, in Virginia, another European artist also began to delineate the natural history of the New World. Mark Catesby, a young naturalist born in Suffolk, was visiting his sister who lived with her doctor husband in Williamsburg. There he began to draw the animals and plants. His subjects were strange and new but at least he had the advantage of seeing them alive, so that he might understand the function of the strange anatomies with which he was faced. One might suppose that he would therefore have avoided the sort of mistakes that led Cassiano's artists to misrepresent the sloth. Alas, Catesby does not seem always to have benefited from his privileged position. He does indeed provide plant settings for his animals. But, almost perversely, often the settings that he gives them are neither accurate from a naturalist's point of view nor even to scale. His flamingo is placed against a branch of so-called black coral, a gorgonian, which only lives in coral reefs, where flamingos never go (fig. 20 and plate 79). Most bafflingly of all, he sets his biggest subject, the magnificent one-ton American bison, against a spray of a *Robinia* tree, drawn to a completely different scale (plate 88).

Catesby was self-taught as an artist and his studies are certainly naïve. But they nonetheless have great freshness and charm. He engraved them himself on copper plates and between 1729 and 1747 published them as *The Natural History of Carolina, Florida and the Bahama Islands.* This was the first in a series of illustrated books on the natural history of North America that culminated in 1827 in perhaps the greatest and certainly the most gigantic of all bird books, entitled *The Birds of America,* by John James Audubon.

Fig. 20
Mark Catesby
A flamingo and gorgonian, c.1725

White-headed Eagle Adult
FALCO LEUCOCEPHALUS.
Fish Catfish, Yellow and Cat

Fig. 21
The white-headed or bald eagle
From John James Audubon,
The Birds of America, 1827–38

Fig. 22
The American flamingo
From John James Audubon,
The Birds of America, 1827–38

Audubon, who from the age of eighteen lived in Pennsylvania where he looked after family property, was obsessed by birds. His quest for them led him to travel ever westwards in search of new species. He hunted them with an unquenchable passion and he drew them with equal enthusiasm. He thought that the standard static profiles that, since ancient Egyptian times, had been the almost universal way of representing birds gave no idea of their vivacity and grace. He determined to draw them in motion. To do so, he took one of his newly shot victims and fixed it on to a board with a squared grid drawn on it. He manipulated the bird's wings and neck into what he considered life-like attitudes and fixed them in position with skewers. The process must have been a fairly blood-spattered one, since his specimens were newly killed.

PLATE CCCCXXXI.

American Flamingo.
PHŒNICOPTERUS RUBER, *Linn.*
Old Male.

1. Profile view of Bill at its greatest extension.
2. Superior front view of upper Mandible.
3. Inferior front view of upper Mandible.
4. Inferior front view of lower Mandible.
5. Interior front view of lower Mandible with the Tongue in

6. Profile view of Tongue.
7. Superior front view of Tongue.
8. Inferior front view of Tongue.
9. Perpendicular front view of the feet fully expanded

PLATE CCLXXI

Frigate Pelican.

TACHYPETES AQUILUS, *Vieill.*

Male Adult.

Fig. 23
The frigate pelican
From John James Audubon,
The Birds of America, 1827-38

The results, however, did indeed bring life to his images. Terns swoop, eagles crouch over their captured prey (fig. 21), and hummingbirds hover in front of flowers.

Audubon also determined that all his subjects should be represented life-size. It was this ambition that led him to print his pictures on the largest sheets of paper available, known as double elephant folio. Huge though the leaves are – 40 by 30 inches (100 × 75 cm) – they were still not big enough to accommodate the biggest of North America's birds in their normal poses. So his flamingo, unlike Catesby's, could not be shown with its neck up but had to be made to bend down so that its head almost touches the ground (fig. 22). This is of little consequence in this instance since that is a pose that the flamingo frequently adopts. But his blue heron has to have its neck bent over its back, as does his Canada goose, which gives them both an awkward and unnatural appearance.

His drawings nonetheless are magnificent. Not only are they very accurate but he had a sense of dramatic design that makes his plates even today the accepted icons of many American species – the wild turkey, the Carolina parakeet (now extinct), the frigatebird in a powered dive (fig. 23) and mockingbirds defending their nest from attack by a rattlesnake. He brought these dramatic drawings to Britain in order to get them engraved for reproduction. They were, however, among the last important natural history drawings to be printed in this way.

A German printer had discovered that a line drawn with a wax pencil on a fine-grained limestone could be inked and printed. He refined the process to such an extent that soon the lithographic process was in use all over Europe. So, unlike Gesner's woodcuts or Audubon's copper-plate engravings, both of which were normally copied from the original drawings by specialised craftsmen, these prints could reproduce the most delicate lines directly from the artist's hand – if he chose, as many did, to draw on the stone. The process led to a new flowering of natural history books. Among the most spectacular were those produced by John Gould.

Fig. 24
Edward Lear
Ramphastos toco
From John Gould,
*A Monograph of the Ramphastidae
or Family of Toucans*, 1834

RAMPHASTOS TOCO. Linn.
Toco Toucan.

The nineteenth century was the heroic age of ornithological discovery. Specimens of new species were flooding into Europe from all over the world. Gould had started his professional career as a taxidermist at the Zoological Society of London but he soon became a full-time publisher. He employed a series of extremely talented bird artists, including his wife Elizabeth; Joseph Wolf, a German bird painter who was particularly skilled in portraying birds of prey; and – perhaps the most talented of all – the eighteen-year-old Edward Lear, who was later to become more famous for his nonsense verse.

Fig. 25
John Gould and William Hart
The golden-winged bird of paradise
From John Gould,
The Birds of New Guinea, 1875–88

DIPHYLLODES CHRYSOPTERA, *Gould*.

Gould's sumptuous folio volumes, in their heavily tooled and gilded full-morocco bindings, appeared in imposing ranks on the shelves of aristocratic libraries throughout Britain. The birds of Britain and Europe, Asia and Australia were surveyed. All known species of hummingbirds, pittas and toucans, trogons and birds of paradise, not to mention two hundred or so marsupials, were each given a separate plate (figs. 24 and 25). In all, Gould's artists figured just one short of three thousand different species.

By the beginning of the twentieth century, it seemed that the age of great scientific

Fig. 26
Henrik Grönvold
*Blue tits, subspecies from
the Canary Islands, c.*1920

natural history painting and the spectacular volumes that required such illustrations was coming to an end. Photography was beginning its reign. At first the process was used as an easy and cheap way of transferring natural history drawings to metal plates or lithographic stones. But then, as photographers became more skilled and their cameras became smaller and more versatile, photographs themselves were used as illustrations.

By the early years of the twentieth century it was possible, if you were sufficiently hardy and a good enough naturalist, to capture an exact image of the creature in front of you with the press of a finger. To some it might have seemed that this would bring an end to the need to catalogue animals by drawing them. But in the event a new need appeared. As increasing numbers of people began to live in towns, cut off from the countryside for much of their lives, and as the countryside itself became increasingly impoverished in variety and reduced in extent, so a new kind of highly expert amateur naturalist arrived who wanted to be able to distinguish the minutiae of a bird's plumage in order to identify its exact species. Few photographs could do that adequately. Thus a new kind of artist appeared, who produced elegant detailed plates which displayed ranks of closely related birds – male, female and juvenile, species, subspecies and race (fig. 26) – to enable dedicated naturalists precisely to identify the creature on which they focused their binoculars.

Now yet another technique for capturing images of the natural world has been invented – the electronic camera. Over the last half century, this has changed from a large unwieldy object the size of a suitcase that produced coarse-grained black and white pictures to a tiny device that can record high-quality colour pictures in light so low that even the human eye has difficulty in perceiving what is in front of it. Optical cables can carry images from underground nest chambers at the end of long narrow tunnels. New vibration-proof mountings, together with long-focus lenses, allow the camera to record head-and-shoulder close-ups of an animal while hovering in a helicopter a thousand feet above it.

You might think that these latest developments would finally bring to an end a tradition that stretches back thirty thousand years. Not so. Today large-scale monographs devoted to particular groups of plants and animals are still produced by artists who welcome the double demands of aesthetic delight and scientific accuracy. And there are still artists who, like Alexander Marshal, paint in private with no intention of displaying their work to the world at large.

And there always will be. For no matter what the ostensible motive for their work, whether it is to lighten the reverential atmosphere of a monastery or to invoke animal spirits in a fertility ritual, to explore anatomy or to catalogue a discovery, there is a common denominator that links all these artists. It is the profound joy felt by all who observe the natural world with a sustained and devoted intensity.

LEON

DO DA VINCI

MARTIN CLAYTON

'ALL THE WORKS OF NATURE WHICH ADORN THE WORLD'

LEONARDO DA VINCI'S studies of the natural world were a central part of his work. He believed that the painter had to understand every facet of creation in order to produce truthful images; and as he perceived these facets as infinitely interlinked, it was impossible to draw firm boundaries between one field of investigation and another. Much of his activity can be related at some level to his projected treatise on painting, the ultimate and unspoken aim of which was, in essence, a complete description of every aspect of natural phenomena. Inevitably, that treatise remained unfinished, but the studies it spawned – into human and animal anatomy, botany, geology, hydraulics, optics, flight, and much else besides – mark Leonardo out as one of the greatest scientists, as well as one of the greatest artists, of the Renaissance.

Leonardo (fig. 27) was born on 15 April 1452 near the town of Vinci, fifteen miles west of Florence in the Arno valley. He was the illegitimate son of a notary, Ser Piero da Vinci, and a peasant girl named Caterina, and was taken into his paternal grandfather's house. Leonardo's early education was basic: he learned to read and write, but his arithmetical skills were always shaky and, though he tried to learn some Latin later in life, he never became comfortable with the language of most scientific writings. Leonardo was left-handed, and throughout his life he habitually wrote his notes in mirror-image, from right to left. This was not an attempt to keep his researches secret, as has been claimed, for his mirror-writing is relatively easy to read with a little practice. Mirror-writing is a common developmental quirk in childhood, and what may have begun as an entertaining trick became a habit that Leonardo never had cause to discard.

Detail of plate 13

By the age of twenty, Leonardo had joined the painters' guild in Florence and was probably working in the studio of the sculptor, painter and architect Andrea del Verrocchio. From the outset of his career Leonardo seems to have been fascinated with the natural world. His earliest dated drawing, of 5 August 1473, is a landscape of the Arno valley, a dense study of rock formations, flowing water and swaying trees. A few years later he made a drawing of a ravine (plate 1) which may appear fanciful at first sight – an impression not helped by the disproportionately large ducks or swans – but which in fact carefully records a type of rock formation found in the upper Arno valley, caused by the erosion of sandstone into tall jagged pillars.

Sheets such as this mark the beginning of Leonardo's obsession with natural processes. He perceived all things – man, animals, plants, watercourses, the earth itself – as dynamic organisms subject to the same laws of nature. Around 1490 he wrote:

Just as man is composed of earth, water, air and fire, so this body of the earth is similar. Whereas man has bones within himself, the supports and framework of the flesh, the world has rocks, the supports of the earth; if man has within himself the lake of blood, wherein the lungs expand and contract in breathing, the body of earth has its ocean, which also expands and contracts every six hours with the breathing of the world; as from the said lake of blood arise the veins, which spread their branches through the human body, likewise the ocean fills the body of the earth with an infinite number of veins of water.

The same cause which moves the humours in every species of animate bodies against the natural law of gravity [literally 'the natural course of their weight'] also propels the water through the veins of the earth wherein it is enclosed, and distributes it through small passages. And as the blood rises from below and pours out through the broken veins of the forehead, as the water rises from the lowest part of the vine to the branches that are cut, so from the lowest depth of the sea the water rises to the summits of the mountains, where, finding the veins broken, it pours out and returns to the bottom of the sea.

Fig. 27
Attributed to Francesco Melzi,
*A portrait of Leonardo da Vinci, c.*1515

Plate 1. A rocky ravine, *c.*1475–80

Plate 2. An outcrop of stratified rock, *c.*1510

Leonardo had an instinctive sense of the vast geological forces that threw up the mountains, raised seabeds to great heights, and changed the form of the continents. He was familiar with the Alps, and recalled climbing Monte Rosa and witnessing the darkening of the blue of the sky at high altitudes. While in the mountains he observed marine fossils (including 'the bones and teeth of fish which some call arrows and others serpents' tongues') lying in strata, and wrote at length about the conflict between the presence of these shells and the biblical account of the Flood. Leonardo's interest in geology must have been well known, for he recalled that 'when I was making the great horse for Milan [see below], a large sackful [of fossil shells] was brought to me in my workshop by certain peasants'.

Leonardo's profound awareness of the endless cycle and unconscionable timescale of these processes is found again and again in his writings, and partly explains the meaning of the distant landscapes of paintings such as the *Mona Lisa* and the *Madonna and Child with St Anne and a lamb* (both Louvre, Paris). He made many drawings of weathered and collapsing rocks, some imaginary, but others, such as plate 2, recording actual formations – in this case an outcrop of splintered, stratified rock bursting out of the earth, drawn with a concentration and acuity not found again before the seventeenth century.

The close observation seen in this drawing was the key to Leonardo's investigations into the natural world. The largest early group of his drawings to survive are studies for the *Adoration of the Magi* (Uffizi, Florence), commissioned in 1481 and left unfinished when Leonardo moved to Milan soon after. For the crowded composition of that painting Leonardo made many drawings of animals – oxen, asses, horses and even, in one perspective study, a camel. Some of these are remarkably down to earth – the gaunt hindquarters of an ass stooping to eat, or an ox standing in solid profile; others are closer to the fantastical spirit of the painting, such as plate 3. The study demonstrates close examination of the anatomy of the horse, with the skin both stretched and bunched over the tensed muscles of the shoulder and neck.

Horses were to be a feature of several of Leonardo's most important projects. A few years after he abandoned the *Adoration*, he was engaged by the ruler of Milan, Ludovico Sforza, to execute an enormous bronze equestrian monument – some three times life size – to Ludovico's father Francesco. Initially Leonardo wanted the horse in a dynamic rearing pose, but the technical demands of the sculpture would probably have been insurmountable. Instead, in 1490, he resorted to a conventional walking or ambling horse with only one foreleg raised, the pose of the few equestrian monuments known from Roman times and of the work's immediate precursors,

Plate 3. A rearing horse, *c.*1480

Donatello's *Gattamelata* and Verrocchio's *Colleoni* monuments. By 1493 the huge clay model was finished, but the cast was never made, for the following year the bronze intended for the horse was requisitioned and sent to Ferrara to make cannon. The project fell into abeyance, and when French forces invaded Milan in 1499 the clay model was used for target practice by French archers and destroyed.

To construct the huge clay model for the mould, Leonardo had undertaken a detailed study of the anatomy of the horse. Most of his research was done in the stables of Galeazzo da Sanseverino,

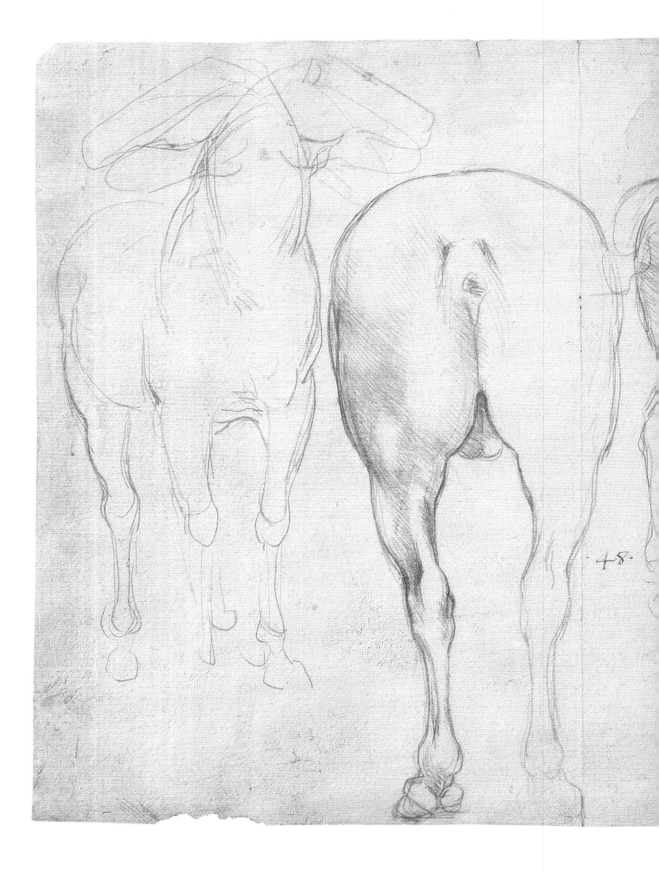

·48·

Plate 4. Studies of a horse, *c*.1490

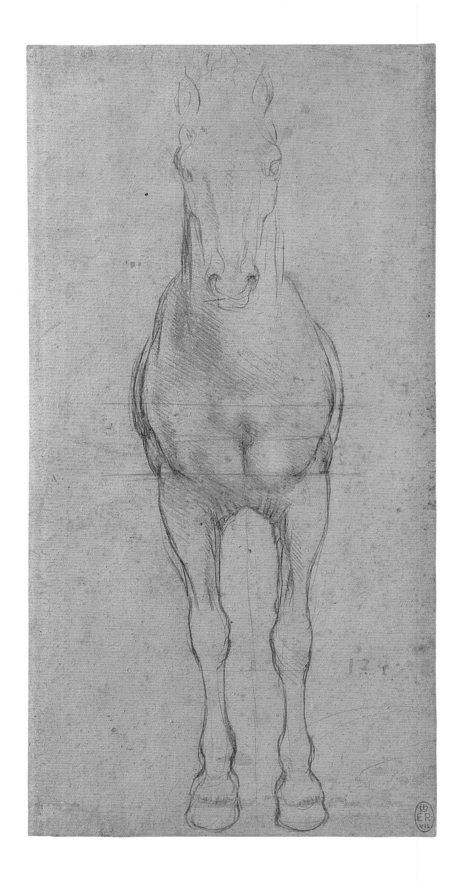

the Captain-General of the Milanese army, and some of his drawings record the exact horse studied, 'the Sicilian' or 'the big jennet of Messer Galeazzo'. A few of Leonardo's studies show the horses in casual poses, such as plate 4, in which the combination of decisive outlines and rapid shading gives a wonderful sense of bulk. Other drawings are more systematic, surveying the measurements and surface modelling of the horse in front and side 'elevations', as if they were architectural studies. Plate 5 shows a large horse in a rather unnerving frontal view, with horizontal lines at the level of its eyes, at three points of its chest and just below its knees. This study was abandoned before Leonardo had annotated it with the relevant measurements; other studies go much further, and record in minute detail the dimensions of the horse's legs, body and head. At the same time, Leonardo seems to have undertaken a study of the internal anatomy of the horse. His early biographer Giorgio Vasari stated that Leonardo composed a manuscript treatise on the anatomy of the horse that was lost when the French invaded Milan, and one surviving drawing shows the viscera of a quadruped, probably a horse.

Notwithstanding the physical difficulties of dissecting such a large animal, Leonardo's work on the horse would have been just one aspect of his extensive investigations into animal anatomy around 1490. The previous decade had seen Leonardo's first significant scientific studies: he had assembled a small library, and begun to record observations on light, colour, perspective and so on. At first his intention was to compile a treatise on the art of painting, but soon he conceived a parallel treatise on the painter's principal subject, the human body. This treatise was to encompass all aspects of man (and woman and child), but whereas the senses, emotions, expressions and so on could be discussed speculatively, anatomy and physiology required material research. Contemporary texts on the functioning of the body barely touched on its physical structure, and Leonardo found himself having to synthesise elements from whatever source he could come by. Some of his drawings, such as a celebrated sheet showing the hemisection of a man and woman in the act of coition, demonstrate complete reliance on traditional beliefs. He was able to obtain a little skeletal human material, and a series of drawings of the skull in various sections is the outstanding achievement of his early anatomical studies. But while human dissection was allowed at the main hospital in Milan, access to fresh cadavers was strictly controlled, and Leonardo seems not to have been able to dissect any human soft tissue at this stage of his career.

Plate 5. A horse from the front, *c*.1490

*In the Middle Ages, brown bears were widely distributed throughout Europe,
including the British Isles. Even in Leonardo's time, they were not uncommon in Italy.
Today, wild Italian bears are restricted to a small population in the Abruzzi
National Park.*

*Bears are the only large mammals that walk, mechanically, in a way comparable
to human beings, that is to say by placing the palm of the foot flat on the ground.
Cats and dogs walk with their heels raised; cows, sheep and horses walk, in effect,
on the tips of a reduced number of toes. The similarity of a bear's gait to that of
humans becomes very clear when the animal rears up on its hind legs as, until recent
times, captive 'dancing' bears were trained to do by wandering
entertainers.*

*The parallels and differences between the
feet of animals must have been very clear to
Leonardo, for he also dissected and drew both
a human foot and a dog's paw. This drawing
shows that he had a thorough understanding
of the mechanics of the foot – the fact that
muscles exert their power by contraction and
that ligaments connecting muscles and bones
operate like ropes and pulleys.* D. A.

Plate 6. The anatomy of a bear's foot, *c.*1485–90

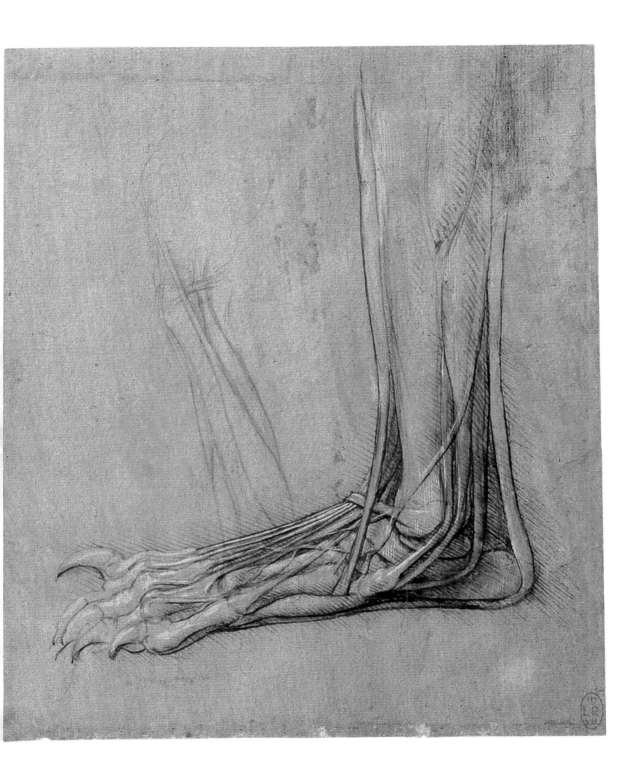

Instead Leonardo resorted to dissecting animals, hoping that this would cast some light on human anatomy, for he always subscribed to the popular belief that 'all terrestrial animals resemble each other as to their limbs … and they do not vary except in length or in thickness'. His subjects included dogs, monkeys, frogs, pigs and bears; sometimes he adjusted the proportions to try to fit the observations to human form, and only anatomical idiosyncrasies reveal which animal was under Leonardo's knife. But where he depicted the anatomy as he found it, he was capable of great objectivity and accuracy. A series of studies of a dissected bear's foot is the finest example (plate 6). In this drawing, Leonardo concentrated on the tendons of the bear's foot and their attachments to the bones. On the upper surface of the foot are the extensor tendons, with the farthest claw raised to demonstrate their action, and the tendons held in place at the angle of the ankle by the anterior annular ligament. This feature clearly impressed Leonardo: twenty years later, in setting out yet another programme of study, he reminded himself to 'make a discourse on the hands of each animal to show in what way they vary; as in the bear, in which the ligament joins the tendons of the toes together over the neck of the foot'.

Leonardo's first wave of anatomical studies subsided after 1490, and he was not to return to the subject for another fifteen years. In 1500 Leonardo had settled back in Florence following the French invasion of Milan and the overthrow of his patron Ludovico Sforza, and around 1503 he began to plan a painting of *Leda and the Swan*. A myth relates that Leda, the wife of Tyndareus, King of Sparta, was seduced by Jupiter in the form of a swan, and subsequently laid eggs from which hatched Helen of Troy, Clytemnestra, and the twins Castor and Pollux.

Leonardo's *Leda* was to have a foreground teeming with plants and flowers, to emphasise the fecundity inherent in the subject-matter. Over the next few years he worked on a series of drawings of plants, many of which are identifiable. The most famous, of a Star of Bethlehem (plate 7), also includes wood anemone and sun spurge, with close attention paid to the peculiar form of the spurge's flower heads. The wood anemone is studied again in plate 8, together with a marsh marigold. And plate 10 shows a stem of the grass known as Job's tears, introduced to Europe from eastern Asia; Leonardo's drawing seems to be the earliest evidence for its presence in Italy.

Plate 7. Star of Bethlehem (*Ornithogalum umbellatum*), wood anemone (*Anemone nemorosa*) and sun spurge (*Euphorbia helioscopia*), c.1505–10

Plate 8. Marsh marigold (*Caltha palustris*) and wood anemone (*Anemone nemorosa*), *c.*1505–10

Several of these botanical drawings are executed in red chalk on orange-red prepared paper, and are as objective as any scientific drawing produced by Leonardo. Plate 11 is a remarkably bold study of a sprig of oak, the dense shading giving it a sense of plasticity and life not found in the work of any contemporary. To the left in that drawing is a more delicate study of dyer's greenweed, separated from the oak by a fold and stitch-holes. A similar fold with identically spaced stitch holes is also found on plate 12, a study of a branch of blackberry, strongly suggesting that the sheets once formed part of a notebook made up from prepared paper.

These drawings were far more detailed than was necessary for the painting of *Leda*, and indicate that Leonardo was now studying botany as an end in itself. Plate 13, of two rushes in seed, is laid out in the same manner as many of Leonardo's anatomical sheets. The accompanying notes read:

This is the flower of the fourth kind of rush, which is the tallest of them, growing three to four *braccia* [1.5–2 metres] high, and near the ground it is one finger thick. It is of clean and simple roundness and beautifully green; and its flowers are somewhat fawn-coloured. Such a rush grows in marshes etc., and the small flowers which hang out of its seeds are yellow.

This is the flower of the third kind or species of rush, and its height is about one *braccio* and its thickness is one third of a finger. But this thickness is triangular, with equal angles, and the colour of the plant and the flowers is the same as in the rush above.

Plate 9. A sprig of blackberry (*Rubus fruticosus*), *c.*1505–10

Plate 10. Job's tears (*Coix lachryma-jobi*), *c*.1510

Plate 11. Oak (*Quercus robur*) and dyer's greenweed (*Genista tinctoria*), *c*.1505–10

Plate 12. A branch of blackberry (*Rubus fruticosus*), *c.*1505–10

Plate 13. The seed-heads of two rushes (*Scirpus lacustris* and *Cyperus* sp.), *c.*1510

Other drawings and comments scattered through Leonardo's papers indicate that he was considering a treatise on botany, concentrating on the physical structure of plants. He had acquired a couple of books on the subject, an '*erbolajo grande*' (large herbal) and an edition of Pietro Crescenzio's *Libro della agricoltura*, both listed in Leonardo's possession in 1504. His 'Manuscript G', in Paris, dating from around 1510, contains sketches and discussions of the growth and branching of trees, such as 'The cherry-tree has the character of the fir tree as regards its branching, which is placed in stages around its main stem ....The elm has the largest branch at the top .... In the walnut tree the leaves which are on this year's shoots are farther apart from each other and more numerous', and so on. The manuscript also includes many passages on the fall of light on leaves, actually more germane to the treatise on painting than to one on botany. A delicate study of a tree (plate 14) – perhaps a young elm – probably dates from around the same time, for the note below reads:

> That part of a tree which is against shadow is all of one tone, and where the density
> of trees and branches is greater, there it is darker because light has less of an impression
> there. But where the branches are against other branches, there the luminous parts show
> themselves brighter, and the leaves shine as the sun illuminates them.

Leonardo's treatise on botany was never written, but some of the plant studies must have been used in the *Leda*, which was painted during the last decade of his life. The original painting was destroyed around 1700, and is known though several copies; these display a variety of plants in different combinations, and it is difficult to reconstruct exactly which of Leonardo's studies found their way into the original.

At the same time that Leonardo began work on the *Leda*, around 1503, he was commissioned by the republican government of Florence to paint a huge mural of the *Battle of Anghiari* in the council chamber of the Palazzo della Signoria. The only portion to be executed was the central scene of the *Fight for the Standard*, a battle of men and horses that led Leonardo to make a comparative study of the expressions of fury in men, horses and lions (lions were kept in a cage behind the Palazzo della Signoria at this time). The *Battle of Anghiari* also led Leonardo to revive his interest in human anatomy, at first concentrating on the superficial aspects, but soon delving below the surface. Leonardo was by now a highly regarded figure in Florentine life, frequently consulted by the government, and his services were used as something of a diplomatic gift. His enhanced status gave him access, for the first time, to human corpses, in the monastery and university hospitals of

Plate 14. A tree, *c*.1510

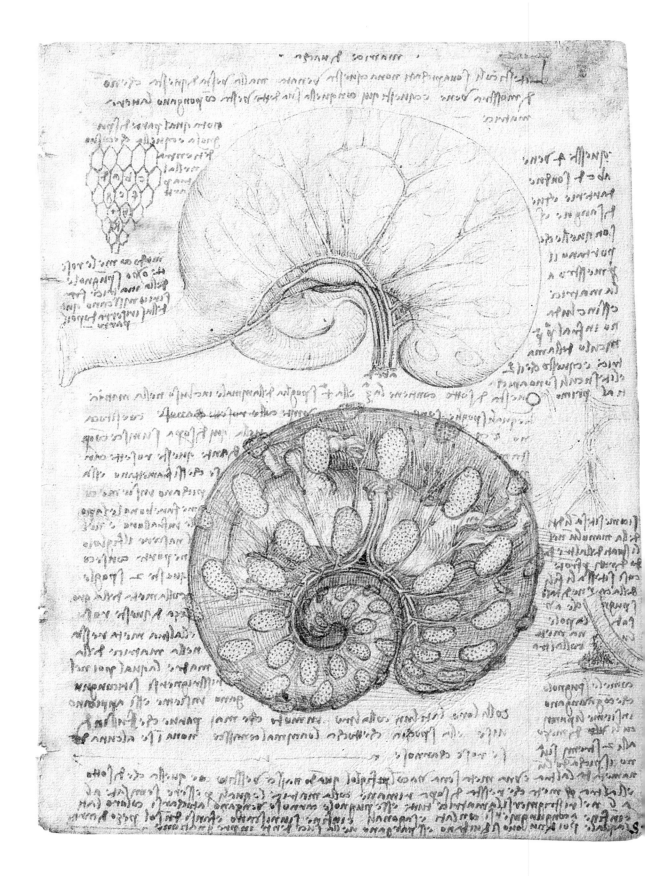

Florence, and later of Milan, Pavia and Rome. Over the next ten years he claimed to have dissected thirty bodies, a number that is not contradicted by the quantity of surviving drawings and notes.

But still Leonardo used animal material on occasion. His astonishing cardiological studies of 1510–13 were based on dissections of the bovine heart. His work towards a projected flying machine, conducted over many years, included a treatise on bird flight and several sensitive investigations of birds' wings. He intended to 'describe the various forms of the intestines of the human species, of apes and suchlike. Then, in what way the leonine species differ, and then the bovine, and finally birds.' He reminded himself to 'describe the tongue of the woodpecker and the jaw of the crocodile' and to 'procure the placenta of a calf when it is born'; and in a small notebook otherwise devoted to human anatomy he made an exquisite study of the uterus of a gravid cow (plate 15).

The upper diagram of plate 15 is an external view of the bicornuate (two-horned) uterus, with the vagina to the left and an ovary at the centre. Below, the uterine wall has been removed, to show the placental cotyledons of the chorion and the foetal calf within the upper 'horn' of the uterus, its head to the left and its legs upwards. Leonardo wrote in the accompanying notes that, on the birth of the calf, the cotyledons united in the manner of the honeycomb sketched at upper left, to be delivered as a single afterbirth. Since he believed the uteri of all animals to be essentially the same, the notes refer not to cow and calf, but to mother and infant, and the results of this investigation were to be applied (mistakenly) to his studies of the human uterus a couple of years later.

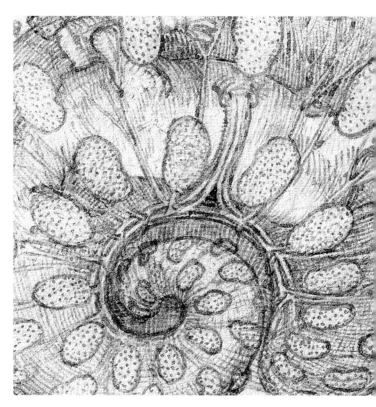

Plate 15. The uterus of a gravid cow, *c.*1508

Perhaps the most disappointing aspect of Leonardo's anatomical researches was his inability to bring his work to a conclusion and organise his material into a form suitable for publication. Though he was in possession of some of the finest anatomical studies ever made, he seems to have put these aside in the last few years of his life, and set off instead in a fresh direction. A memorandum in one of Leonardo's notebooks of *c.*1513–14 sets out a programme of study of movement, ending, 'Write a separate treatise describing the movements of animals with four feet, among which is man, who likewise in his infancy crawls on all fours'.

A charming sheet of animal studies (plate 16) includes highly naturalistic drawings of cats sleeping and grooming, more stylised depictions of cats fighting and lionesses stalking, and a rather unexpected dragon. The fragmentary note at the bottom of the page gives an indication of Leonardo's main interest: 'Of flexion and extension. This animal species, of which the lion is prince because of its spinal column which is flexible.' An analogous note appears on a related sheet of perhaps a couple of years later (plate 17), which concentrates instead on the horse but also includes a lion and several depictions of St George fighting the dragon: 'The serpent-like movement

Detail of plate 16

is the principal action in animals and is double, the first occurring lengthways and the second crossways'.

It is unlikely that Leonardo took these studies very far – indeed it is hard to know how much mileage there could have been in such a subject – but his interest in the movement of animals found expression in one of his final projects. In late 1516 Leonardo moved from Rome to the court of the young French king, Francis I, in the Loire valley. He seems to have started work on a project for another equestrian monument, probably to the king himself, and in addition to several sheets of sketches of the whole monument there survive a number of studies of the horse in motion. As we have seen, Leonardo had studied the horse in great detail at various points in his career, but, not content with his earlier drawings – which he had retained as part of his workshop resources – he began a campaign of study of the movement of the horse. Most of the drawings examine horses' legs from various angles, either planted firmly on the ground or raised in a walking or ambling gait. Plate 18 is one of these drawings, with the horse seen in front and back views reminiscent of some of the studies for the Sforza monument (such as plate 5) of almost thirty years before. The accompanying, rather opaque note begins, 'This movement made by the forequarters of the horse is divided in two parts, the first of which consists of raising the right part more than the left'.

Like so many of Leonardo's projects, this late equestrian monument seems never to have left the drawing board. Leonardo died at Amboise on 2 May 1519, and the contents of his studio were inherited by his two faithful companions Francesco Melzi and Gian Giacomo Caprotti, known as Salaì. Melzi took the drawings and manuscripts back to his family villa near Milan, and spent many years trying to put them in some sort of order, transcribing passages from Leonardo's notes in an attempt to compile the *Treatise on Painting* that the artist himself had never completed.

Detail of plate 17

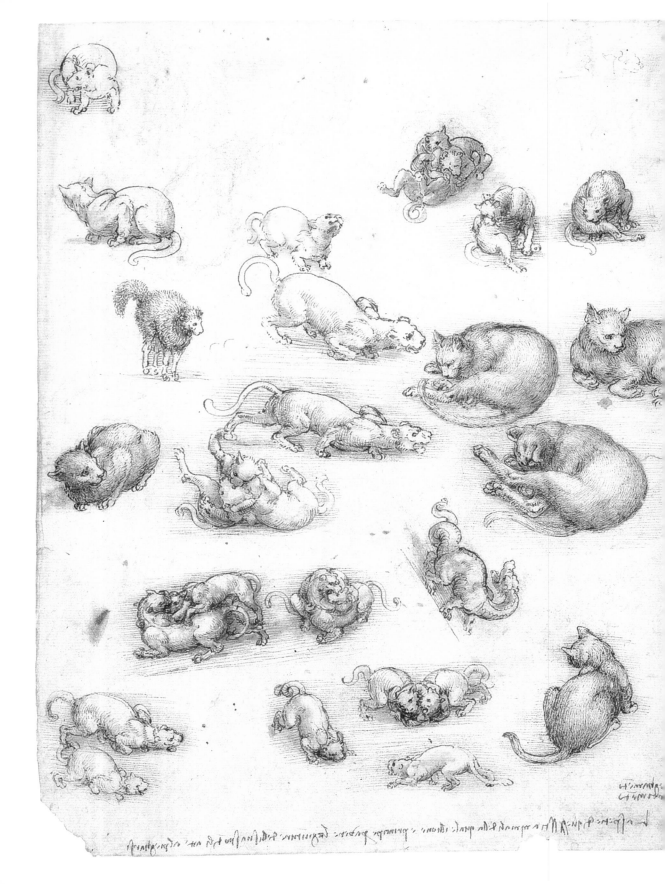

The reality of dragons was not doubted. The first great encyclopaedias of natural history, by Gesner and Aldrovandi, catalogued them in considerable detail, even though these works were published nearly a century after Leonardo's death.

There were thought to be several different kinds. Some were without legs. They, doubtless, were inspired by stories of giant snakes told by travellers returning from Africa and the tropics of the Far East. The encyclopaedists made it clear that legless dragons were even more formidable than the largest snakes by depicting them wearing small coronets. Other dragons, however, had legs. The number varied. One kind had a single pair, just behind the head. Others had four legs or even eight. And one spectacular monster, the hydra, had seven heads.

This sheet of drawings vividly records Leonardo's musings on their nature. He clearly set out to record the precise postures of cats and lions as they stalk and play, threaten and wrestle. And then, while doing so, he wondered how a quadrupedal dragon with the skeleton and muscles analogous to those of a cat would crane a rather longer neck or twirl a more extended tail. D. A.

Plate 16. Cats, lions and a dragon, *c.*1513–16

Several kinds of dragon were thought to have possessed wings sprouting from the shoulders. Leonardo, as an experienced and extremely knowledgeable comparative anatomist, would have pondered the bodily mechanisms that such creatures must have possessed to operate such limbs. On this sheet of studies of a saint on horseback who might tackle such monsters he does indeed give his dragon a pair of wings – and the practical comparative anatomist in him very sensibly makes them a modified version of a pair of forelegs, like the wings of a bird, rather than a more improbable extra set of limbs operated from the shoulders.   D. A.

Plate 17. Horses, Saint George and the Dragon, and a lion, *c*.1517–18

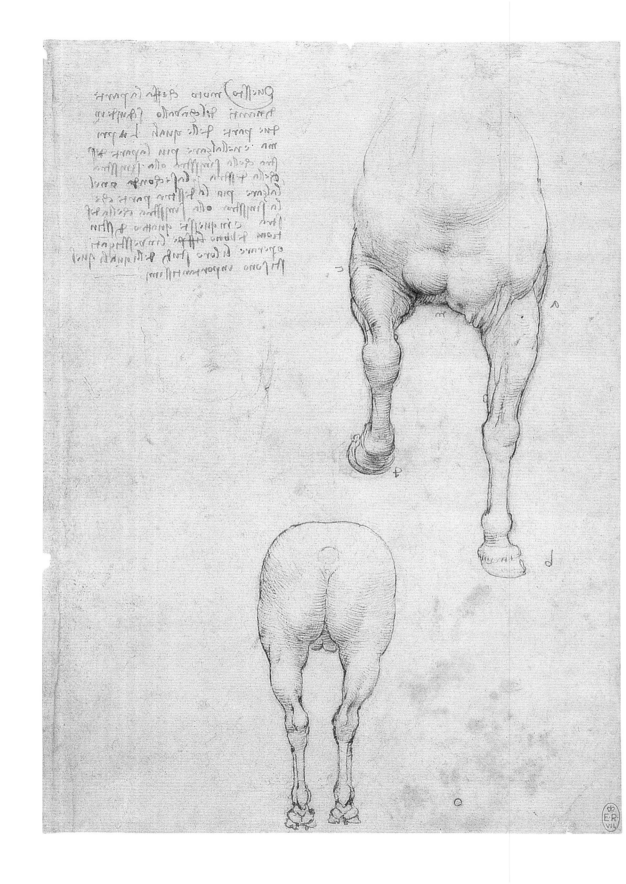

Plate 18. The chest and hindquarters of a horse, *c.*1517–18

(A version of Melzi's compilation was finally printed, through the agency of Cassiano dal Pozzo, in 1651.) After Melzi's death around 1570, his son sold Leonardo's papers to the sculptor Pompeo Leoni, who pasted the loose drawings into the pages of several albums, of which two survive – the Codex Atlanticus, now in the Biblioteca Ambrosiana in Milan, which consists mainly of technical drawings, and the volume that contained all the drawings now at Windsor (the binding has been preserved as an empty shell).

The present Windsor volume was auctioned after Leoni's death in Madrid in 1608, and by 1630 it was in England, in the possession of the great collector Thomas Howard, 2nd Earl of Arundel. During the Civil War Arundel left England for the Low Countries and eventually for Italy, but it is not known whether he took the Leonardo album with him. It is not recorded again until 1690, when it was seen in London in the possession of William III and Queen Mary. The means by which the volume entered the Royal Collection is unknown, but it is probable that it was acquired, by purchase or gift, by Charles II (reigned 1660-85).

Detail of plate 16

TH
CASSIA

APER MUSEUM OF
DAL POZZO

REA ALEXANDRATOS

'WITH THE TRUE
EYE OF A LYNX'

THE 'PAPER MUSEUM' of the antiquarian and collector Cassiano dal Pozzo (1588-1657) was a visual encyclopaedia of the ancient and natural worlds consisting of thousands of drawings and prints. The subject matter covered the surviving remains of Roman civilisation (including architecture, wall paintings, mosaics, reliefs, inscriptions and household objects), Renaissance architecture, artefacts connected with the early Christian church, maps, religious processions and festivals, costumes, portraits, as well as every aspect of the natural world, from birds, fishes and other animals to plants, fungi and fossils. Of roughly 7,000 surviving drawings from the Paper Museum (taking no account of the prints), around 2,500 are of natural history subjects.

Cassiano dal Pozzo (fig. 28) was born in Turin and received most of his education in Pisa, where his father's cousin Carlo Antonio dal Pozzo was archbishop and advisor to Ferdinand I, Grand Duke of Tuscany. He graduated in civil and ecclesiastical law in 1607 and served briefly as a judge in Siena in 1612 before moving to Rome, where he remained for the rest of his life. In 1623 he entered the household of Cardinal Francesco Barberini, nephew of the newly elected pope, Urban VIII (reigned 1623-44), accompanying the cardinal on diplomatic missions to France and Spain in 1625-6. Cassiano was appointed *maestro di camera* (head of the household) in 1633 and, although he never commanded great wealth, this and other official positions provided him with the influence and income to participate in the cultural pursuits for which the Barberini papacy (and the cardinal's circle in particular) became well known.

With his younger brother Carlo Antonio dal Pozzo (1606-1689) – himself an enthusiastic collector and 'a youth most curious of natural things' – Cassiano lived in a small *palazzo* on the via dei Chiavari in Rome which housed his famous collection of paintings, prints and drawings, as well as books, natural specimens, scientific instruments, enclosures with live birds and animals, and a laboratory for chemical experiments and dissections. When Cassiano died in 1657, his friend Carlo Dati wrote a eulogy in which great claims were made for him as a naturalist: 'Not content with the simple description and history of nature', Dati asserted, Cassiano 'went further, with the true eye of a lynx, to study its very anatomy'. Dati was here referring to Cassiano's membership of Europe's first modern scientific academy, the Accademia dei Lincei, which took its name from the lynx, an animal fabled for its sharp vision.

The Accademia dei Lincei was established in Rome by Prince Federico Cesi in 1603, half a century before either the Royal Society in London or the Académie des Sciences in Paris. It numbered Galileo amongst its members and placed great emphasis on observation as the key to unravelling the mysteries of nature and of the universe. Visual documentation played a vital role in

Fig. 28
Pietro Anichini
A portrait of Cassiano dal Pozzo
Frontispiece to Carlo Dati's funerary
oration, *Delle lodi del commendatore
Cassiano dal Pozzo*, 1664

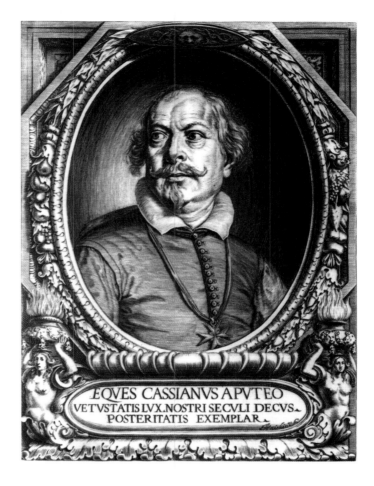

EQVES CASSIANVS APVTEO
VETVSTATIS LVX.NOSTRI SECVLI DECVS.
POSTERITATIS EXEMPLAR

this enterprise, and it is this documentary (rather than artistic) impulse that accounts for the large corpus of natural history drawings assembled by Cassiano.

Cassiano's presentation piece to the academy on his election in 1622 was a book by Giovanni Pietro Olina entitled *L'Uccelliera* (The Aviary). He had helped the author to assemble the material and wrote to Cesi that the publication was 'an experiment, to see if with a little expense and dedication, the drawings I am assembling could be used as illustrations to such writings'. Although the *Uccelliera* was a reworking of a popular manual on the art of fowling (Antonio Valli da Todi's *Canto degli augelli* of 1601), Cassiano added new plates based on drawings 'very diligently made by Vincenzo Leonardi'. Cassiano himself was not an artist; he commissioned artists to make drawings directly from objects or specimens, and to copy others from existing drawn sources. That we know

so little about the artists he employed is, however, symptomatic of the documentary bias of the Paper Museum, the subjects depicted being more important than the artists involved. While Cassiano is well known for his patronage of leading artists such as Nicolas Poussin and Pietro da Cortona, those whom he employed for the Paper Museum were more obscure. Only one sheet is signed, a drawing of a sparrow by Vincenzo Leonardi (*fl.*1621–46), and his is the only name that can be definitely associated with the natural history drawings. Very little is known about Leonardi except that he accompanied Cassiano on his trip to France in 1625 and provided the illustrations for two of the three natural history publications associated with the Paper Museum, the *Uccelliera* and Giovanni Battista Ferrari's *Hesperides* of 1646, a treatise on the cultivation of citrus fruit. Many of Cassiano's drawings of citrus fruit (such as plates 19 and 20) can therefore be attributed to Leonardi, and on stylistic grounds many of the other natural history drawings in the Paper Museum have also been assigned to the artist.

As well as commissioning drawings, Cassiano acquired large groups of earlier drawings, including the '*libri dipinti*' (painted books) of the Accademia dei Lincei, bought by Cassiano from Cesi's widow following her husband's death in 1630. Among these were more than a hundred drawings of fossil woods which later provided the models for etchings in the 1637 *Trattato del Legno Fossile Minerale* (Treatise on Fossilised Mineral Wood) by the *linceo* Francesco Stelluti. Fossil woods had fascinated Cesi because they seemed to combine the properties of more than one kingdom, being of an 'intermediate nature between plants and minerals' (as he wrote to Cardinal Barberini). An even larger body of Lincean drawings, preserved today in eight volumes in the library of the Institut de France in Paris, was dedicated to ferns, mosses and especially fungi (of which there are more than five hundred folios), considered 'imperfect plants' because they seemed to lack reproductive structures such as flowers or seeds.

Plate 19. Pummelo (*Citrus grandis*): whole and half fruit, *c.*1640

Fig. 29
Mattheus Greuter
Melissographia, 1625

Notes on colour, smell, taste, weight, season and the locality in which the specimens were found can be seen on many of these botanical and mycological drawings, including the proud inscription '*microscopio observatio*' next to enlarged details. These are among the earliest illustrations ever produced with the revolutionary instrument that had been developed by Galileo and presented to his fellow Linceans in 1624. With this 'aid to the eyes', wrote the physician and *linceo* Johannes Faber, 'our Prince Cesi saw to it that many plants hitherto believed by botanists to be lacking in seeds were drawn on paper'. The first printed illustration made with the microscope appeared in 1625, an engraving entitled *Melissographia* which included three magnified views of a bee and its anatomical details (fig. 29); the bee was an emblem of the Barberini family and of Pope Urban VIII, to whom the plate was dedicated by the academy.

Cassiano himself followed up the *Uccelliera* of 1622 with a number of extraordinary *discorsi* or short manuscript treatises on ornithology (the subject that seems to have engaged him most deeply). Three of these survive, on the bearded vulture, the ruby-throated hummingbird and the Dalmatian and European pelicans; a further *discorso*, on the toucan he saw in the collection of Louis XIII in Paris in 1625, was used by Faber in the Lincean publication *Animalia Mexicana* (1628), but is lost. The *discorso* on the pelicans is the only one for which the accompanying drawings survive (including plates 21 and 22). The pelicans had been shot in marshes near Ostia and brought to Cassiano in 1635. His *discorso* carefully distinguished between the two species – the Dalmatian pelican was 'slightly smaller', and differed from the great white or European pelican in the colouring of its plumage, which was 'ash-coloured' rather than a 'dirty white'. Cassiano's intention of illustrating works on the natural sciences with high-quality images taken from life found shape in these *discorsi*. Their detailed observations were meant to be read together with the drawings: the hummingbirds were 'of the precise size seen in the painted figures';

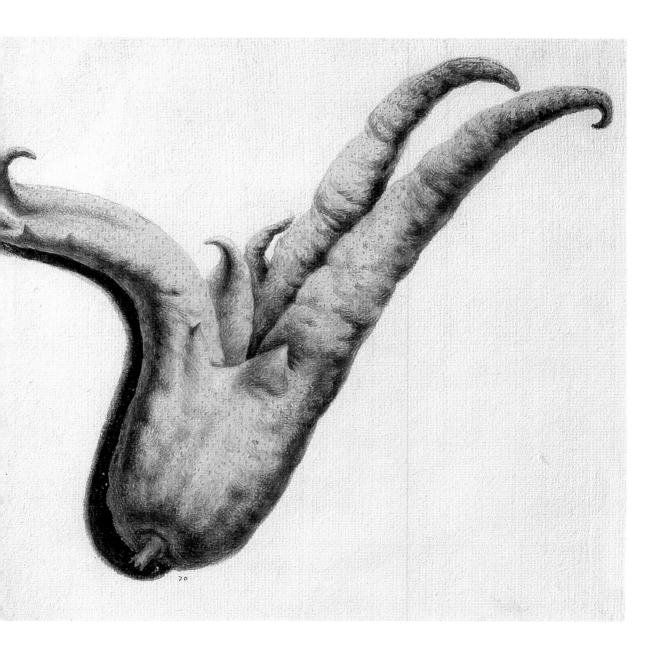

Plate 20. Digitated lemon (*Citrus limon*), *c*.1640

Plate 21. European pelican (*Pelecanus onocrotalus*), 1635

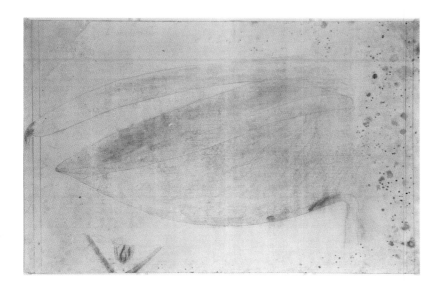

Reverse of plate 22

and the syrinx of the European pelican (the 'forked organ of the throat', which Cassiano was able to see by opening the bird's pouch and pointing it towards the light) was 'drawn in chalk on the back of the painting' and can be seen on the reverse of the drawing reproduced as plate 22.

A striking feature of the *discorsi* is the attention paid to subtleties and variations of colour. Of the beak and pouch of the European pelican, which had been drawn actual size (plate 22), Cassiano noted, 'The lower part of the beak, to which was attached the pouch or crop, was a beautiful indigo colour which is, however, somewhat less vivid in this drawing; the upper part was so beautiful that it was wonderful to see because the colours were in waves or bands, flowing into each other – flesh pink, yellow and an azure blue which was almost a pale indigo'. The pouch was a 'beautiful straw-coloured yellow' when the bird first arrived (barely 'a few hours after it had died of shot wounds'), but it soon began to darken, and on measuring its capacity Cassiano found it could contain 'fourteen pounds of water easily'. He noted that the Bishop of Cervia had reared his own pair of European pelicans by feeding them on fish (they consumed 20–25 pounds, or 9–11kg between them each day) as well as on bread, onions, green vegetables and tripe. Three years later, when another European pelican was brought to Cassiano, he had it dissected and discovered that it was female because of the eggs it carried.

OVERLEAF:
Plate 22. Head of a European pelican (*Pelecanus onocrotalus*), 1635

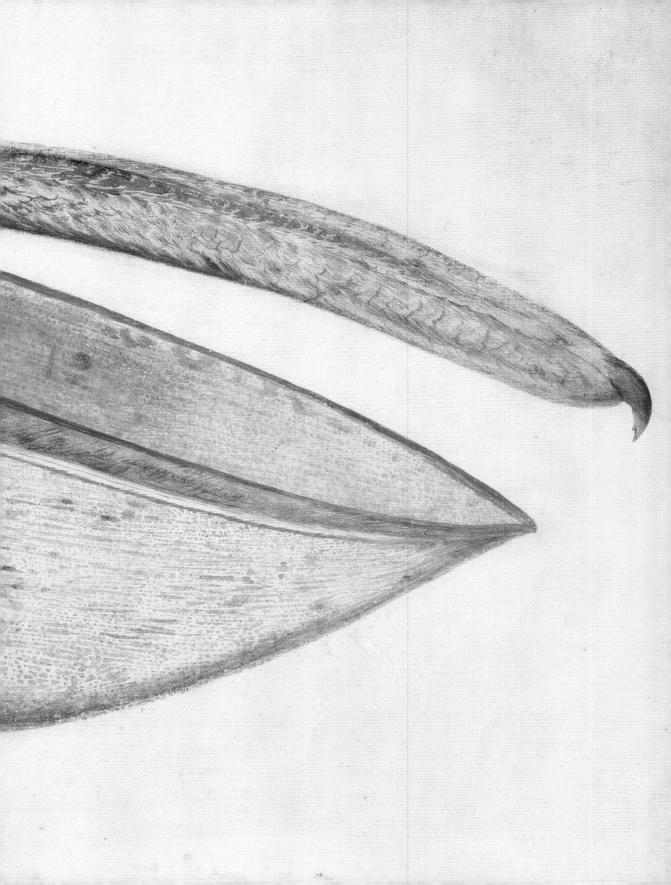

Plate 23. African civet (*Civetticus civetta*), *c*.1630

The African civet is a native of sub-Saharan Africa. It is largely solitary and marks its territory by smearing rocks and tree trunks with powerfully smelling secretions from anal glands. These secretions were, in Cassiano's time, harvested from captive specimens for use in the manufacture of perfumes and were highly valued. It is not surprising therefore that the artist has given some prominence to these glands in his drawing.

Edward Topsell, Cassiano's English contemporary, writes in his Historie of Four-footed Beastes *(1607) that the secretion 'must be taken away every second or thirde day or else the beast doth rub it forth of his own accord upon some postes in his kennel, if he be tamed and inclosed. One ounce of it if it be pure and not sophisticated, is sold for eight Crowns at the least. There be impostors, who do adulterate it with an Oxes gall, Styrax, and Honey. This is of a strange savour, and preferred before Musk by many degrees, yet it smelleth worse if it be held to the nose.'*

I believe that the last statement, at least, is still true!　　　　　　　D. A.

Cassiano's artist cannot have seen the living animal. Three-toed sloths have a highly specialised diet of leaves from only a few species of tree, so it would not have survived the long voyage from the New World to Europe. Even today, few zoos manage to keep them alive.

Apart from this, however, there is ample evidence in the illustration that the original draughtsman was working from a dead specimen. The animal spends its life hanging from trees and is able to do so without muscular effort because its claws are hooked. Indeed its legs are so minimally muscled that it would not have the strength to assume, let along maintain, the posture shown. If detached from its branch and put on the ground its legs splay laterally and the animal is only able to move by a kind of swimming action.

Cassiano's artist has shown, accurately, that the sloth's coarse hair grows thickly around the base of its claws and that it has no bare soles or pads there. But it is evident that, if it normally assumed the posture shown, that hair – and indeed the claws – would quickly be worn away.

The maned sloth is the rarest of the three species of three-toed sloths and is today officially listed as endangered. It is found in the coastal forests of eastern Brazil. D. A.

Plate 24. Maned three-toed sloth (*Bradypus torquatus*), 1626

OVERLEAF:
Plate 25. Common dolphin (*Delphinus delphis*), *c.*1630–40

336

173

Plate 26. Leg and feather of a white stork (*Ciconia ciconia*), *c*.1630–40

Plate 27. Head of a white stork (*Ciconia ciconia*), *c*.1630–40

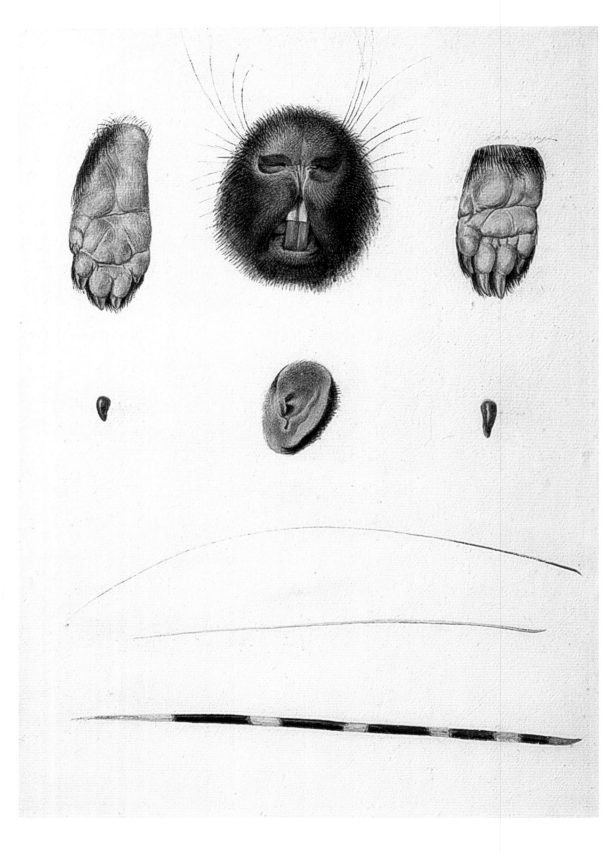

The circulation of Cassiano's manuscript *discorsi* may have been limited, but the Paper Museum was made available to scholars from all over Europe, and was widely used in the seventeenth century as an instrument of study and research. The drawings also formed the basis of lively exchanges between Cassiano and his correspondents in Italy and elsewhere, including the French antiquarian Nicolas-Claude Fabri de Peiresc, Peter Paul Rubens and Fabio Chigi (the future Pope Alexander VII). Drawings and specimens were sent as gifts between friends: both Chigi and Cardinal Barberini had their own private collections of exotic animals, and over the years Chigi sent Barberini gazelles, a civet cat from Africa and guinea fowl from Malta. As the cardinal's man for all 'animals out of the ordinary that turn up' (according to a contemporary's account), Cassiano was able to commission drawings of animals in Barberini's zoo, such as the civet cat (plate 23). The animal was prized for the musk extracted from its anal glands (the anal region is highlighted in the drawing) and had been discussed at length by Faber in his *Animalia Mexicana*. A specimen was dissected for Cassiano in his laboratory and was the subject of a treatise bearing a dedication to him by the physician and botanist Pietro Castelli (*Hyaena odorifera*, 1638).

The civet cat was among many animals represented in the Paper Museum mentioned in a letter from Cassiano to Peiresc in 1634, together with an elephant, gazelles, a hippopotamus, a roebuck, a mongoose, a jerboa and a sloth. The drawing of the sloth (plate 24) was not made directly from a specimen but was probably copied from a picture hanging on a staircase of the Escorial, which Cassiano recorded in his 1626 diary of the papal legation to Spain: 'a portrait of the lazy or tardy animal from Brazil, called ironically *Perrico Ligero* [the name given to an agile member of the weasel family] … whose hair is curly, between tan coloured and grey, with very tall legs at the front and considerably shorter ones behind, of slender body, beaver teeth and small eyes'. Although most of Cassiano's drawings are exceptional for their accuracy and detail, this is by no means the only item in the Paper Museum that relies on secondary sources rather than on first-hand observation; at least four of Cassiano's depictions of fish are based on drawings produced almost a century earlier for Ippolito Salviani's *Historiae naturalis de piscibus* of 1554.

'Salvianus's fishes done to the life' were among the many drawings seen by Philip Skippon during a visit to the dal Pozzo collection in 1662, after Cassiano's death, along with one of 'a dolphin brought to the fishmarket in Rome, having one fin on the middle of the back, a pair

Plate 28. Anatomical details of the common or crested porcupine (*Hystrix cristata*), *c*.1630–40

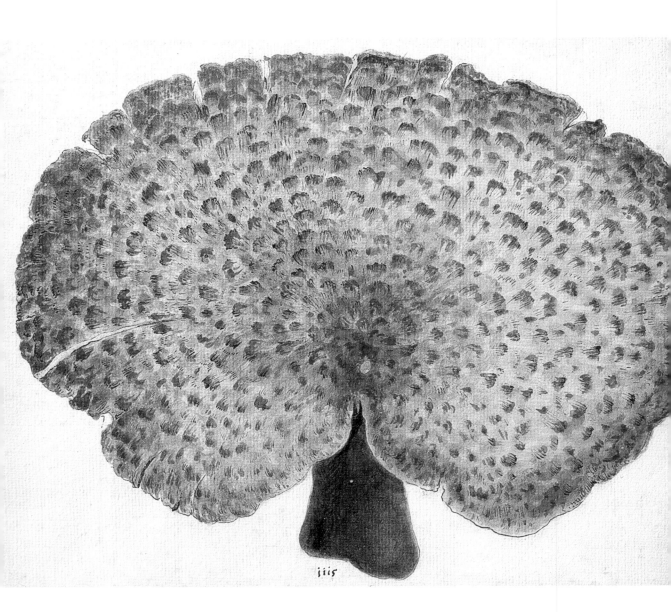

Plate 29. Dryad's saddle (*Polyporus squamosus*), seen from above, *c.*1650

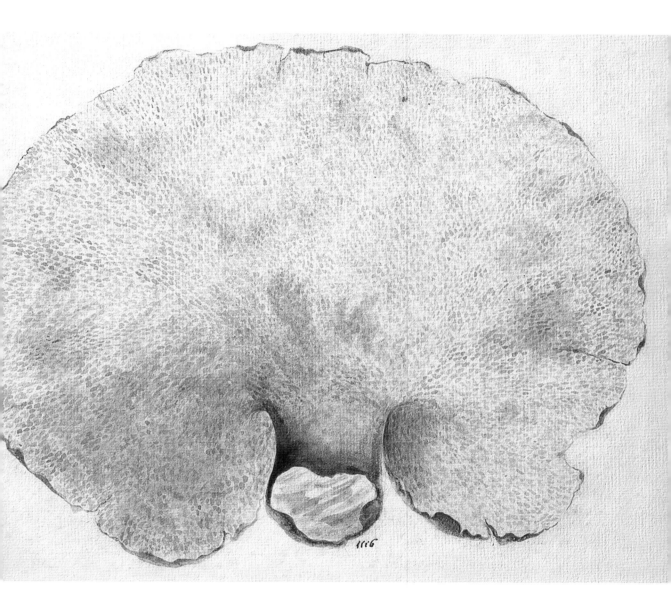

Plate 30. Dryad's saddle (*Polyporus squamosus*), seen from below, *c*.1650

Plate 31. Noble pen shell (*Pinna nobilis*), *c.*1630–40

of fins under the gills, a longish snout, wide mouth, a forked tail, and well armed with sharp teeth' (plate 25). Dolphins were a common sight in Italian coastal waters and would have been considered fish (rather than mammals) in Cassiano's day. Skippon went on to note that the drawings he saw were 'painted very exactly, the heads, legs and other parts of animals being distinctly drawn'. The attention to anatomical detail is indeed a distinctive feature of Cassiano's drawings: while the whole specimen might be drawn on a reduced scale to fit the sheet, the details, such as the head, leg and feather of a stork (plates 26 and 27) or the snout, claws, paws, ears and quill of a porcupine (plate 28), were depicted life size. Different views of the same specimen might be shown, as in the drawings of the bracket fungus known as dryad's saddle (plates 29 and 30), seen from above and below; or of the noble pen shell (plate 31), depicted both closed and open in order to show its internal anatomy, including the fibrous threads that could be woven into a fine golden textile. Botanical and mycological studies illustrated different stages of growth – a lily's transformation from flower to fruit, or the development of a red cage fungus from young unopened stage to fully expanded fruitbody; fruit was commonly depicted in sections and segments, as in drawings of a pummelo (plate 19) and a deformed melon (plate 32).

Such freak specimens were a prominent feature of the curiosity cabinets or *Wunderkammern* of the day, and had populated the sheets of the Paper Museum's sixteenth-century predecessors, the most illustrious being the encyclopaedic collections of Ulisse Aldrovandi in Bologna and Conrad Gesner in Zurich (see pages 17–19). The English scientist and philosopher Francis Bacon, whose work was greatly admired by Cassiano for its 'advancement of thought in all the sciences' ('if he were not living in England', Cassiano wrote to a fellow Linceo, 'I would want to make every effort to make him one of us'), urged natural philosophers to collect 'deviating instances, such as the errors of nature, or strange and monstrous objects in which nature deviates and turns from her ordinary course'. A drawing in Cassiano's collection of an amphisbaena, a two-headed snake described by Pliny, was the source of a woodcut published by Faber (fig. 30), who wrote: 'Just as I became convinced that the two-headed amphisbaena was probably the stuff of myth and fable rather than of truth, the Cavaliere Cassiano dal Pozzo, one of our Linceans, showed me the most truthful image of an amphisbaena in the form of a drawing with all the appropriate colours'. Cassiano did not focus exclusively on the exotic, rare or bizarre, but he did share an interest in abnormality. 'While we are

Lat. Pinna clausa.
Romæ Tromleone di mare

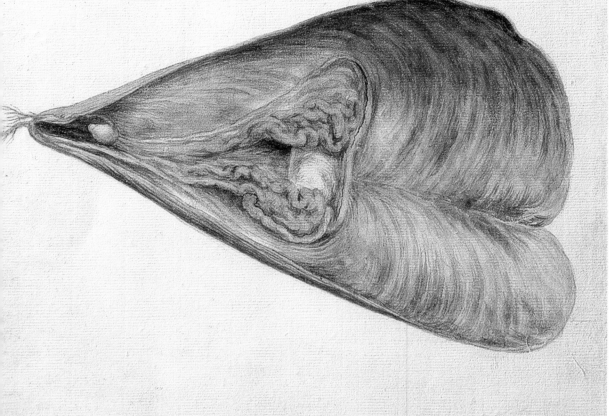

Pinna eadem aperta

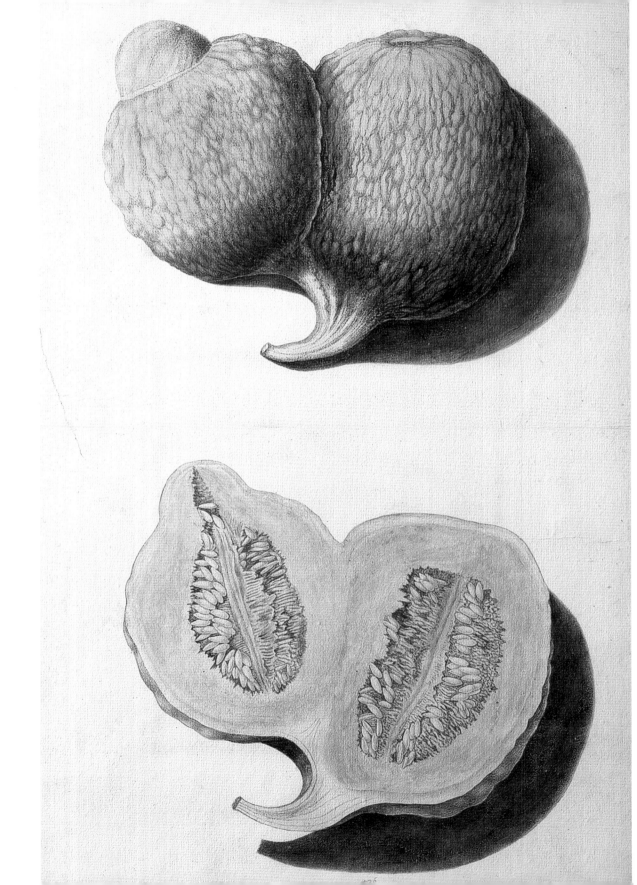

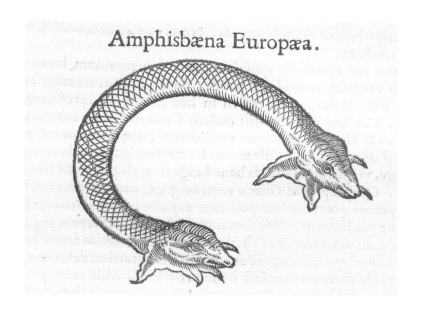

Fig. 30
Amphisbaena Europaea,
Illustration from Johannes
Faber's essay on the animals
of Mexico in Francisco
Hernández, *Rerum medicarum
Novae Hispaniae thesaurus,*
1651

generally horrified by monstrosities in the case of human beings, we love them in fruit', wrote
Ferrari in the *Hesperides,* where several of the illustrations are based on Cassiano's drawings of
misshapen citrus fruits, such as the digitated lemon (plate 20). The deformity in such specimens is
caused by the action of a mite on the bud of the flower, but Ferrari explained their 'fingered'
appearance by inventing an elaborate poetic narrative involving the tragic transformation of a
mythical youth into a citrus tree.

Also typical of curiosity cabinets was the assembly of gems, minerals and semi-precious stones
depicted in plate 33, one of a group of drawings in which the striking line-up of specimens evokes
the way they were stored in these collections. Most of the agates, jaspers and jades depicted on this
sheet were believed to have magical curative properties; others were prized for their rarity or exotic
origin (the green and white jade on the top row for example, numbered *56,* which probably came

OVERLEAF:

Plate 33. Gems, stones and amulets, *c.*1630

Plate 34. Fruits, seeds and legumes, *c.*1630

Plate 32. Deformed melon (*Cucumis melo*), *c.*1630–40

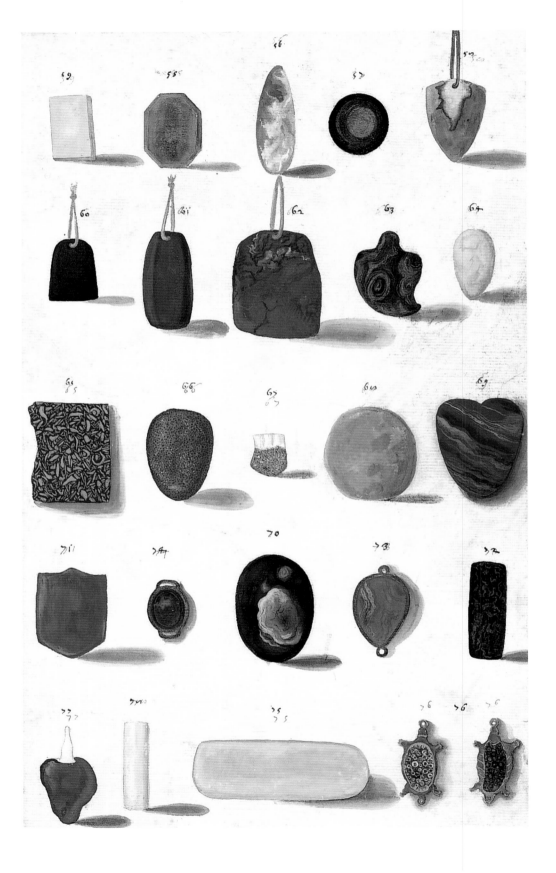

419

Plate 35. Peony (*Paeonia mascula*), with the root of *Paeonia officinalis*, *c*.1610–20

Plate 36. Deformed broccoli (*Brassica oleracea,* var. *italica*), *c*.1650

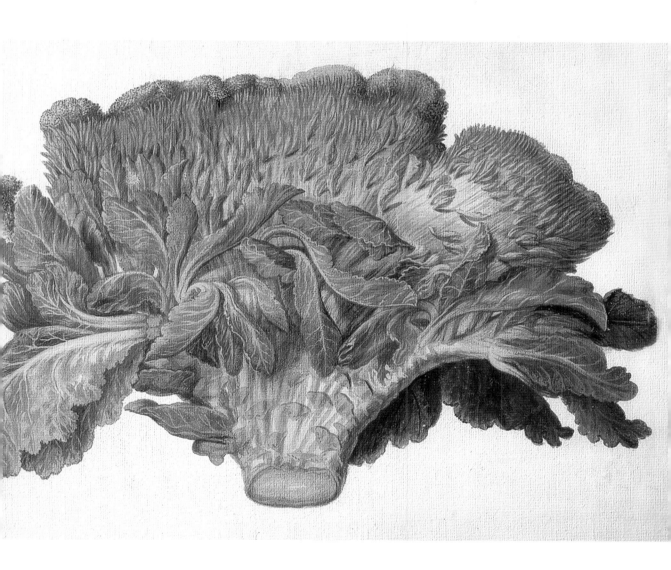

Detail of plate 33

from China), as well as for their pharmaceutical properties. The Abur stone from Rajasthan depicted at centre left (numbered *65*) was one of a number of specimens that have been identified with items in Cesi's geological collection.

Exotic fruits, seeds and legumes from Africa, Asia and America – including a tropical raphia palm nut on the top row – are displayed in a similar line-up in plate 34. Like many Europeans, the Linceans were fascinated by the ever increasing variety of previously unknown species that were daily being reported or imported from the Americas. In 1626 Cassiano had a copy made of an Aztec herbal that had been presented to Cardinal Barberini in Spain, and together with his fellow Linceans Cassiano spent many years editing for publication the monumental work on Mexican flora and fauna by Philip II's physician Francisco Hernández, the *Rerum medicarum Novae Hispaniae thesaurus*. Expeditions like the one undertaken by Hernández to the Americas between 1570 and 1577 were gradually undermining the authority of ancient writers on natural history that had been so dominant in Leonardo da Vinci's day; for in describing hundreds of plants and animals for which no match could be found in the work of the ancients, Hernández and others opened the way to systematic doubt and to the new emphasis on empirical enquiry and sensory verification characteristic of the 'new science'. An important part in this development was played by specialised collections of natural specimens (precursors of natural history museums) and archives of images like those of Cassiano and the Lincei, who attempted to classify the natural world through visual description.

Cassiano's natural history drawings are usually marked with small inked arabic numerals whose precise function remains unclear. Different views of the same specimen usually bear the same number, but these views may also be numbered consecutively (like the dryad's saddle, plates 29 and 30, numbered *1115* and *1116*). The sequences jump around on a group of drawings illustrating selections of gems, stones, minerals and fossils (including plate 33), in which the same numbers are sometimes assigned to several specimens on the same sheet. Cassiano's numbering, it may be surmised, corresponds to lost inventories of the collection rather than to an all-encompassing classificatory scheme.

Carlo Antonio inherited and continued to add to the collection after his brother's death in 1657, and it was his grandson Cosimo Antonio dal Pozzo (1684–1740) who sold the dal Pozzo library, including the Paper Museum, to Pope Clement XI in 1703. In 1714 the purchase was completed by the pope's nephew, Alessandro Albani, from whom in turn most of the Paper Museum (along with many other drawings from the Albani collection) was acquired by George III in 1762. Few of the 'precious books' extolled by Dati ('in which are most vividly drawn and clearly recorded so many animals of the air, of the earth and of the sea, the most beautiful works of Nature') were kept intact; the majority were split up and rebound or integrated into other parts of the Royal Collection, and many of the natural history drawings were sold in the early twentieth century. Nonetheless, Cassiano's Paper Museum remains one of the most impressive manifestations of the new spirit of empirical investigation that transformed the study of natural history in the seventeenth century.

ALEXA

ER MARSHAL

SUSAN OWENS

'HIS CURIOUS
BOOKE OF FLOWERS
IN MINIATURE'

ALEXANDER MARSHAL (*c.*1620–1682) painted his exquisite florilegium, or flower-book, over a period of some thirty years; he probably began it in around 1650, and he was still adding to it in the year of his death. Its 159 sheets show the plants and flowers of English gardens over the course of a year, from the crocuses and fritillaries of early spring to autumnal gourds and Chinese lanterns. The numerous 'plant portraits' in the florilegium are remarkable for their delicacy, accuracy and beauty. Although many manuscript florilegia were produced on the Continent during the seventeenth century, Marshal's flower-book appears to be unique in English art of the period.

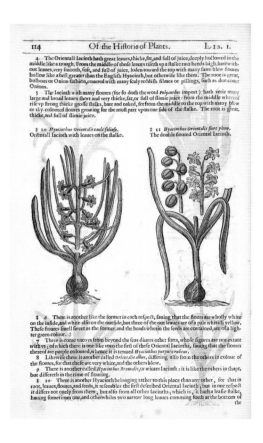

Fig. 31
Hyacinths
From John Gerard, *The Herball, or Generall Historie of Plantes*, first published 1597, enlarged and amended by Thomas Johnson, 1636

Marshal lived during a time of exponential growth in the varieties of plants cultivated in England. Since the late sixteenth century specimens had begun to flood in from all over the world, particularly the Near East. These were often highly scented and dramatic. Tulips, hyacinths, narcissi, turban ranunculi, lilies and the spectacular crown imperial, which was first introduced to a European garden from Turkey in the 1570s, caused astonishment and delight when they emerged from bulbs and tubers. Such novelties led to passionate enthusiasms. It was the era of so-called 'tulipomania', centred in Holland in the 1630s, when tulip bulbs were highly prized and fetched enormous sums. During this time of rapid change the cultivation of gardens became the absorbing pursuit of an elite of scholars and collectors, by whom they were increasingly regarded as extensions of their collections – outdoor cabinets of curiosities. The first botanical garden in Britain, the Oxford Physic Garden, opened in 1621. It was a time also of great activity in the construction of grand houses with large formal gardens to be planted. New and exotic species were procured in the 1610s for the gardens of Hatfield House in Hertfordshire, the seat of Robert Cecil, 1st Earl of Salisbury, and for Wilton House in Wiltshire, the gardens of which were designed in the 1630s for Philip Sidney, 4th Earl of Pembroke. The widespread

Fig. 32
Hyacinths
From Basilius Besler,
Hortus Eystettensis, 1613

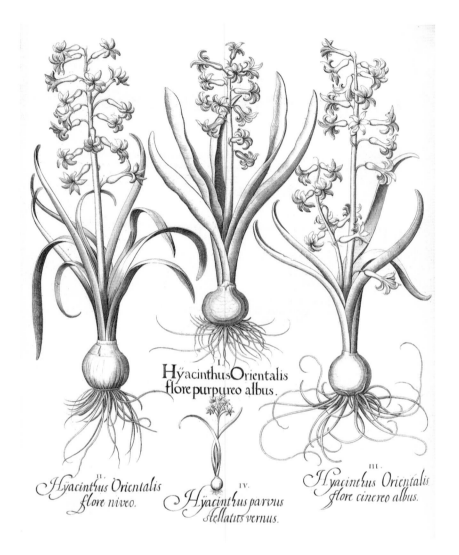

passion for plant novelties, and the lengths to which enthusiasts would go to indulge it, were satirised by Andrew Marvell in his poem of *c*.1650-52 'The Mower, Against Gardens', in which the narrator drily observes: 'Another world was searched, through oceans new, / To find the *Marvel of Peru*'. However, the new attitude contributed greatly to the study of the natural world and its ordering and classification.

As a volume devoted to the depiction of beautiful flowers, Marshal's florilegium belongs to a new era of natural history illustration which had developed in parallel with the enlivened garden

Fig. 33
Title page of John Parkinson, *Paradisi in sole: paradisus terrestris*, 1629

culture. Previously the illustration of plants had mostly been limited to herbals, which were illustrated with comparatively crude woodcuts printed on the same page as the text (fig. 31). In the early seventeenth century, when plants were increasingly cultivated and valued for their beauty rather than for their medicinal properties, botanical art had begun to reflect the prevailing mood by presenting flowers in an aesthetic rather than a diagrammatic manner. Pictures of plants began to take precedence over text and to occupy more space, usually a page to themselves. Etching and engraving, which permitted much more detail, began to replace the woodcut. One of the most sophisticated of the new kind of printed books was the great two-volume *Hortus Eystettensis* of Basilius Besler (1561-1629), which was published in Nuremberg in 1613. Its 374 plates, engraved with great virtuosity by at least six different printmakers, portray plants from the famous garden of Johann Conrad von Gemmingen, Bishop of Eichstatt (fig. 32). In London in 1629 the apothecary and gardener John Parkinson (1567-1650) published an important horticultural treatise *Paradisi in sole: paradisus terrestris* (the punning title can be translated as 'the earthly paradise of Park-in-sun'), which is subtitled 'A garden of all sorts of pleasant flowers …' (fig. 33). This was the first English book devoted to the beauty of flowers as distinct from their medicinal use. Indeed, at the beginning of Parkinson's later work devoted to the medicinal properties of plants, the *Theatrum Botanicum* of 1640, he

Plate 37. Seville orange (*Citrus aurantium*), purple crocuses (*Crocus vernus*), grass snake (*Natrix natrix*) and caterpillar of goat moth (*Cossus cossus*), *c*.1650–82

Plate 38. Purple crocuses (*Crocus vernus*), cloth of gold crocus (*Crocus susianus*), liverwort, double form (*Hepatica nobilis*), poppy anemones (*Anemone coronaria*) and jay (*Garrulus glandarius*), *c*.1650−82

Plate 39. Hyacinths (*Hyacinthus orientalis*), Persian iris (*Iris persica*), Spanish daffodils
(*Narcissus hispanicus*) and purple crocus (*Crocus vernus*), *c.*1650−82

Plate 40. Narcissus (*Narcissus radiiflorus*), crown imperial (*Fritillaria imperialis*), poet's narcissus (*Narcissus poeticus*) and auriculas (*Primula × pubescens*), *c*.1650–82

notes ruefully: 'From a Paradise of pleasant Flowers, I am fallen (*Adam* like) to a world of profitable Herbes and Plants …'.

Marshal's florilegium was not, like many works in the present book, made as a document of scientific discoveries, but for pleasure. Nor was it intended for publication, or for sale (apparently Marshal claimed to have been offered 300 pieces of gold for it, but would not accept the offer), but rather to be enjoyed privately and shown to friends and acquaintances. Nevertheless, it too is a record of new discoveries, as it depicts the results of horticultural triumphs in the raising of recent imports, along with native plants such as rosebay willow herb and heart's ease. Among these plant portraits are the most fashionable and exotic flowers of the time – the crown imperial, varieties of auricula and hyacinth, broken tulips and the Marvel of Peru, all of which were recent introductions to English gardens.

Little is known about Marshal's life, and the date of his birth is unrecorded. However, in addition to his activity as an artist, it seems that he was equally known to his contemporaries for his large collection of insect and bird specimens, and for his expertise on horticultural matters. Marshal was acquainted with the most renowned gardeners of his day, including the younger John Tradescant, Henry Compton, Bishop of London, and John Evelyn. The educationalist and Prussian émigré Samuel Hartlib described him in his diary as 'one of the greatest Florists' (i.e. plant-growers), and as a dealer in roots, plants and seeds from the Indies and elsewhere. Hartlib also records that Marshal was 'a Merchant by profession' who had 'lived for some years in France speaking French perfectly', a period which probably coincided with his youth.

Plate 41. Auriculas (*Primula × pubescens*), *c.*1650−82

Plate 42. Tulips (*Tulipa gesneriana*), *c.*1650−82

Plate 43. Mourning iris (*Iris susiana*), germander speedwell (*Veronica chamaedrys*), broad-bodied chaser dragonfly (*Libellula depressa*), turban ranunculus, double form (*Ranunculus asiaticus*), flesh fly (possibly *Sarcophaga carnaria*), honeywort (*Cerinthe major*) and bigroot cranesbill (*Geranium macrorrhizum*), c.1650−82

William Freind, Marshal's great-nephew and eventual heir, records that he was a gentleman with 'an independent fortune [who] painted merely for his amusement'. That Marshal was, in his time, both well-known and well-regarded as an artist is evident from his inclusion in Sir William Sanderson's treatise *Graphice* (1658), as one of 'our Modern Masters comparable with any now beyond seas', specifically as a specialist in depicting the natural world: 'Then have we *Marshall* for *Flowers* and *Fruits*'.

Detail of plate 49

The earliest record of Marshal's life dates from 1641, when it seems that he was staying in the house of John Tradescant the Younger (1608–62) at South Lambeth in London. The elder John Tradescant (c.1570–1638) and his son were renowned collectors, gardeners and plant-hunters, credited with introducing a great many new botanical species to England. Their house, which became known as 'Tradescant's Ark', became the first museum in the country open to the public, and the Tradescants' immense cabinet of curiosities eventually formed the foundation collection of the Ashmolean Museum in Oxford. The Ark was a place where, as one visitor remarked, one could almost be persuaded that 'a Man might in one daye behold and collecte into one place more Curiosities than hee should see if hee spent all his life in Travell'. One of Marshal's activities during the period at which he was residing with the younger Tradescant was the creation of a florilegium on vellum (now untraced) devoted to the 'divers outlandish herbes and flowers' growing in the gardens at South Lambeth. The catalogue of the Ark's

us flore
alici,

Plate 44. Common German flags (*Iris germanica*), Montpellier ranunculus (*Ranunculus monspeliacus*), turban ranunculus (*Ranunculus asiaticus*) and bluebells (*Hyacinthioides non-scripta*), *c*.1650−82

Plate 45. Biternate peony (*Paeonia mascula*), turban ranunculus (*Ranunculus asiaticus*),
tulip (*Tulipa gesneriana*) and peony (*Paeonia officinalis*), *c*.1650–82

Plate 46. Velvet rose (*Rosa gallica*), unidentified rose (*Rosa* sp.), flaxleaf pimpernel (*Anagallis monelli*), damask rose (*Rosa damascena*), water forget-me-not (*Myosotis scorpioides*) and Austrian copper rose (*Rosa foetida*), *c.*1650–82

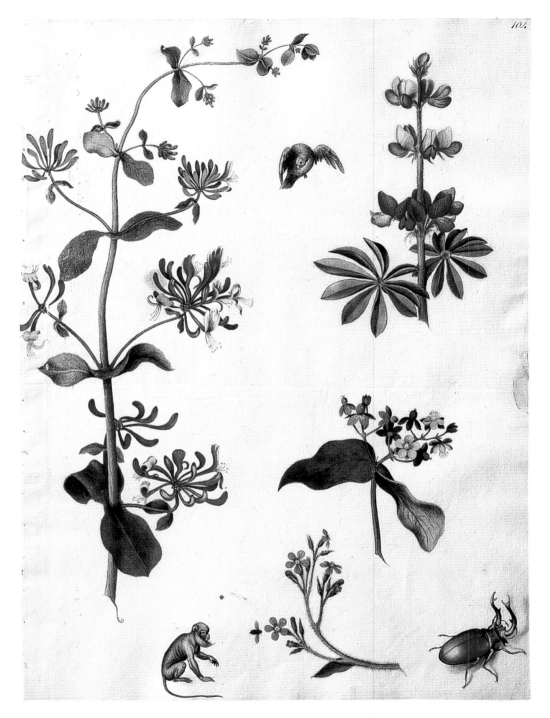

Plate 47. Honeysuckle (*Lonicera periclymenum*), African grey parrot (*Psittacus erithacus*), hairy lupin (*Lupinus hirsutus*), tutsan (*Hypericum androsaemum*), black howler monkey (*Alouatta pallia*), green blowfly (*Lucilia* sp.), alkanet (*Pentaglottis sempervirens*) and stag beetle (*Lucanus cervus*), *c.*1650−82

contents, the *Musaeum Tradescantianum*, which was published by the younger Tradescant in 1656, includes a reference to 'A Booke of MR. TRADESCANT'S choicest Flowers and Plants, exquisitely limned in vellum, by Mr. *Alex: Marshall'*. Marshal had completed this work by 1650, when Hartlib noted in his commonplace book, 'John Tradesken hath a booke [by Marshall] very lively representing most of the th[ings] hee hath [i.e. in his garden]'. Marshal also made drawings of a number of insects in Tradescant's collection; an inscription on one such sheet (part of an album now in the Academy of Natural Sciences, Philadelphia) reads, 'This day butterflye I drew, by one that was in John Tradescant Closet, hee tould me himselfe that he cought it in virginia …' (fig. 34).

By 1653 Marshal seems to have completed a significant part of his own florilegium. This was in part, as Hartlib noted, 'a Picturary of his owne Herbary' (i.e. plants from his own garden), although,

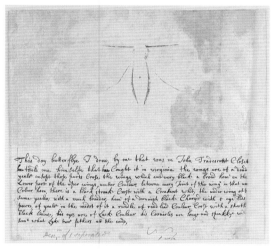

Fig. 34
Alexander Marshal
Recto: *A lime butterfly and notes*, c.1650
Verso: *Notes on an unidentified butterfly*, c.1650

Plate 48. Common sunflower (*Helianthus annuus*) and greyhound (*Canis familiaris*), c.1650–82

Although Marshal devotes the great majority of his album to flowers, he seems unable to resist, every now and then, drawing animals that particularly delighted him.

He was a dedicated collector of insects. Travellers from all over the world brought back specimens for his collection, some of which were undoubtedly very spectacular. Those he chose to illustrate in his album, however, are all British species and comparatively modest in their appearance. On this page he shows the caterpillar of the swallowtail butterfly feeding on one of its favoured food plants, fennel, and also the rather duller pupa that it becomes before turning into the winged adult. He does not show the much more spectacular adult butterfly. Perhaps he preferred to draw living creatures rather than dead ones.

The blue and yellow macaw must surely have been a pet, as was his greyhound which he drew repeatedly. The dragonfly he portrays could also have been alive: it is the southern hawker, a species still common today in southern England. Like all dragonflies it habitually perches with its gauzy wings outstretched as if inviting portraiture and it regularly returns to the same perch as it flies in circuits catching mosquitoes and other small insects. So it would not have been too difficult for Marshal to have repeatedly observed the living creature in exactly the same position.

The unidentifiable bird, on the other hand, is shown in such a strange posture that it seems more likely to have been a dead stuffed specimen from a collector's cabinet; and the crayfish, from its colour, was surely boiled. D. A.

Plate 49. Blue and yellow macaws (*Ara ararauna*), southern hawker (*Aeshna cyanea*), wasp (family *Vespidae*), unidentified bird, caterpillar and chrysalis of swallowtail butterfly (*Papilio machaon*), native or white-clawed crayfish (*Austropotamobius pallipes*), greyhounds (*Canis familiaris*), leaf of ivy-leaved cyclamen or sowbread (*Cyclamen hederifolium*) and caterpillar of oak eggar moth (*Lasiocampa quercus*), c.1650–82

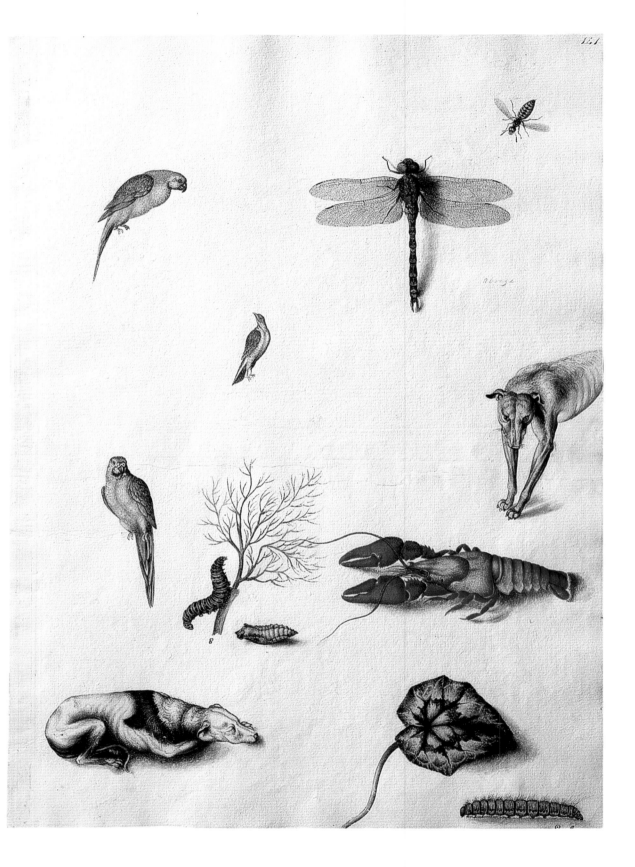

as many of the plants which he drew are included in the plant catalogue of the *Musaeum Tradescantianum*, he probably also used models from Tradescant's garden at South Lambeth. By this time, following a spell in Ham to the west of London, Marshal was living in Islington, where he had lodgings and a garden, and he also had a residence near Lincoln's Inn Fields. He is recorded as having been involved in the importation of plants during this time, and also as having been employed by the Earl of Northampton either in his garden at Castle Ashby or at Canonbury Manor, his Islington residence. During the 1650s and 1660s Marshal also produced a number of paintings. These were usually copies from the works of other artists; his flower studies, however, including the exquisite *Flowers in a Delft jar* (fig. 35), were evidently made from first-hand observation.

Marshal's other great horticulturalist friend was Henry Compton (1632–1713), a younger brother of the Earl of Northampton. In 1667 Marshal moved from London to take up an appointment as Steward to the Hospital of St Cross, outside Winchester, where Compton was Master. Not long after this Compton was appointed Canon of Christ Church, Oxford; in 1674 he became Bishop of Oxford and in 1675 he was made Bishop of London, whereupon he took up residence at Fulham Palace. At this time Marshal returned to London and joined Compton's household, an arrangement which continued even after Marshal's marriage in 1678.

Bishop Compton was devoted to the garden he cultivated at Fulham Palace, where he is recorded as having 'above 1000 species of Exotick Plants in his stoves and Gardens, in which last place he had endenizon'd a great many that have been formerly thought too tender for this cold climate'. It is recorded that 'few days a year [passed] but he was actually in his garden ordering and directing the removal and replacement of his trees and plants'. Compton kept up a large correspondence with botanists in Europe and America, by which means he introduced many previously unknown plants to England. He was also helped by chaplains in overseas posts, who sent him seeds and who could sometimes be prevailed upon to collect insect specimens for Marshal as well. A number of the plants which Marshal included in his florilegium grew in Compton's garden;

Plate 50. Passion flower (*Passiflora incarnata*) with ladybirds (*Coccinella septempunctata*) and caterpillar of unidentified moth or butterfly, autumn crocus, variegated form (*Colchicum variegatum*), leaf of ivy-leaved cyclamen or sowbread (*Cyclamen hederifolium*), humble plant (*Mimosa pudica*), caterpillar of unidentified moth, and caterpillar of cockchafer (*Melolontha melolontha*), *c.*1650–82

Fig. 35
Alexander Marshal
Flowers in a Delft jar, c.1663

a note in Marshal's hand on the verso of folio 146 (plate 52), which depicts a ginger plant, reads 'the trew figure of Ginger as it grew att Fulham'.

Marshal evidently added to his florilegium over the course of many years. On 1 August 1682 John Evelyn, a central figure of Restoration gardening and author of *Sylva, or a Discourse on Forest-Trees* (1664) noted in his diary: '& thence to *Fulham* to visit the *Bish*: of *London*, & review againe the additions which *Mr Marshall* had made of his curious booke of flowers in miniature, and Collection of Insecta'.

Annotations mostly in Marshal's hand on the back of each sheet of the florilegium record the name of the flower or creature depicted. Provenance is occasionally noted, and this information attests to Marshal's wide circle of friends and colleagues who would give, exchange and trade in botanical specimens. His drawing of a Guernsey lily, perhaps the earliest depiction of the flower, is accompanied by the note 'sent to me by Generall Lambert august 29 1659 fro[m] Wimbleton'. Lambert was a Parliamentary general during the Civil War, who after a quarrel with Cromwell in 1657 retired to his beloved garden. His enthusiasm for tulips made him the butt of jokes, and he was satirised by a contemporary poet as 'the Knight of the Golden Tulip'.

One of the great charms of Marshal's florilegium is the variety of compositional devices employed, to express the individual nature of each plant. His sheet of auricula studies (plate 41) shows the plants neatly arranged, each occupying its separate place on the page, prefiguring the later vogue for auricula 'theatres' in which the plants would be arranged on a succession of platforms in front of a painted screen.

Detail of plate 41

OVERLEAF:

Plate 51. Red-billed toucan (*Ramphastos tucanus*), pomegranate, double form (*Punica granatum*), common crane (*Grus grus*), meadow saffron or autumn crocus (*Colchicum autumnale*), possibly caterpillar of buff-tip moth (*Phalera bucephala*), vine branch (*Vitis vinifera*), scarlet macaw (*Ara macao*), mona monkey (*Cercopithecus mona*), filbert or hazel nuts (*Corylus maxima* or *Corylus avellana*), caterpillar of cabbage white butterfly (*Pieris brassicae*) and common frog (*Rana temporaria*), c.1650−82

Parrots of one kind or another have been kept as pets by Europeans since classical times. Their ability to imitate words, their amiable way of clambering around their cages and their undemanding taste in food made them popular throughout history. One of Alexander the Great's generals brought back what seems to have been a green ring-necked parakeet from his Indian campaigns in 323 BC and grey parrots, with their particularly marked gift for mimicking human speech, have been brought to Europe from Africa since even before medieval times. Marshal shows African greys elsewhere in his album, on both occasions in active and accurate poses.

By the mid-fifteenth century parrots from the New World had arrived in Europe. Marshal, judging from his album, seems to have known, perhaps even possessed, two of the most spectacular species, the scarlet macaw (on the sheet opposite) and the blue and yellow macaw that he drew twice on another sheet (plate 49).
Both have a wide distribution in South America, ranging from Panama in the north down to southern Brazil. In Marshal's time they were common in the lowland coastal forests, and so easily acquired by European seafarers. They are robust creatures and will survive on a diet of nuts and seeds such as could easily be provided throughout a long voyage. So within a few years of Europe's discovery of the New World macaws became much treasured and spectacular pets.

Toucans are also South American birds. Marshal shows a rather doleful-looking individual. Perhaps it was a little unwell when he drew it, for its hunched posture is not characteristic of the bird. They are easily tamed and will hop around on a meal table, sipping from glasses of water and taking food from a diner's plate. They feed primarily on fruit and will delicately pick up a cherry or some other morsel with the tip of their huge beak, throw it into the air and catch it in their throat. Such amiable behaviour, however, is not fully representative of the bird's nature. It is also a meat-eater and will pluck the young from another bird's nest with equal enthusiasm D.A.

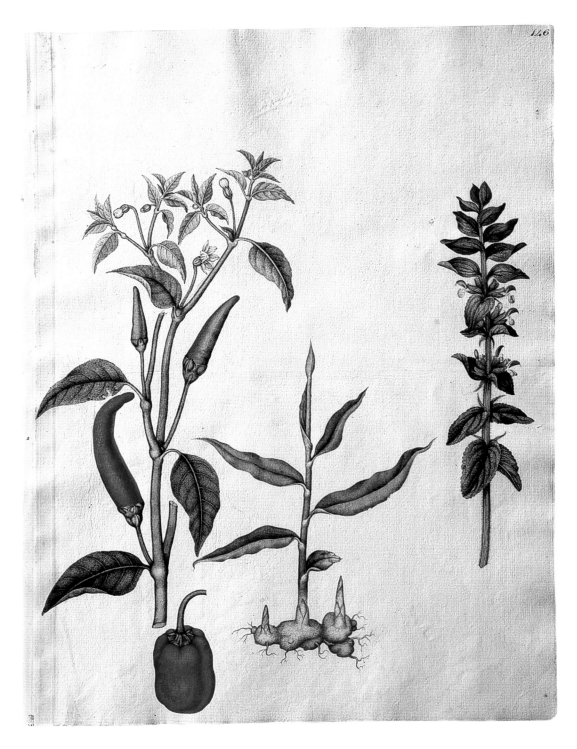

Plate 52. Chilli pepper (*Capsicum frutescens*), sweet pepper (*Capsicum annuum*), ginger (*Zingiber officinalis*) and clary sage (*Salvia horminum*), *c*.1675–82

Plate 53. Joseph's coat (*Amaranthus tricolor*) and white-flowered bottle-gourd (*Lagenaria siceraria*), *c.*1650−82

In contrast to this orderliness, Marshal's passion flower (plate 50) flies off the page in exotic exuberance, while below it the humble plant sits modestly in its simple earthenware pot. Other compositions are strikingly dramatic, for example plate 38, in which a dead jay lies heavily at the bottom of the sheet otherwise occupied by delicate crocuses, and is treated in a *trompe-l'oeil* manner in contrast with the unshaded flowers above. On other sheets miniature drawings of birds, dogs and other creatures are included, perhaps merely as *jeux d'esprit*. In plate 48 a greyhound appears to shelter underneath the extravagantly curving petals of a giant sunflower, and a black howler monkey makes a surprising appearance in the bottom corner of plate 47. No fewer than thirty-eight different species of insects appear throughout the florilegium. As one

Detail of plate 52

might expect from a renowned collector of insect specimens, these are depicted in more detail and with greater accuracy than the other animals (see particularly the superb dragonfly in plate 49).

The physician Christopher Merret wrote of Marshal: 'I know an Ingenious gentleman, who this way hath made all his colours for plants, which he hath drawn to the life in a large volume of the most beautiful flours of all sorts in their proper and genuine colour' (*The Art of Glass*, 1662). The colours of the florilegium remain remarkably fresh and bright to this day. Marshal described his experiments with pigments, reporting that he extracted them from 'flowers, or berries, or gums or roots'. He remarked that 'the Searche of Colours has Cost me much time in finding out' and rightly predicted that those which he used in the florilegium 'will bee as fresh a hundred years hence'.

Marshal died at Fulham Palace in December 1682. After the death of his wife in 1711 the florilegium, along with Marshal's other works, passed to their eldest nephew, Dr Robert Freind, headmaster of Westminster School, and in due course it became the property of his son Dr William Freind. After his death it left the family and was sold, passing through the hands of a Mr Way. In 1818 it was purchased in Brussels by a Ross Donnelly, who gave it to his friend John Mangles. Mangles had the florilegium re-bound, in two volumes, and presented it to George IV, although the date and circumstances of the gift are not recorded. The florilegium has since been disbound.

The inscription on Marshal's tombstone in Fulham parish church (broken when the church was rebuilt in 1880) included the lines: 'PROLEM NON RELIQUIT AT PROBITATE ET INGENIO LONGIOR HUIC FACTA EST DATA VITA FUIT' (He left no issue, but by reason of his integrity and gifts he will live longer than the life which was vouchsafed him).

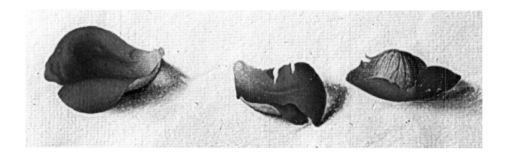

MARIA SI

LLA MERIAN

SUSAN OWENS

'GREAT DILIGENCE, GRACE AND SPIRIT'

MARIA SIBYLLA MERIAN (1647-1717) was exceptional both as an artist and as a naturalist. Her pioneering expedition between 1699 and 1701 to the Dutch colony of Surinam in South America, where she studied the insects indigenous to the country, resulted in her magnificent work of 1705, *Metamorphosis insectorum Surinamensium* (The transformation of the insects of Surinam). This volume in imperial folio, comprising sixty engraved plates showing the life-cycles of butterflies and other insects with textual commentaries, was one of the most important works of natural history of its era as well as the first scientific work to be devoted to Surinam. Of the ninety-five watercolours on vellum by Merian in the Royal Collection, sixty are related to the published plates of the *Metamorphosis*. The remaining thirty-five are still lifes depicting flowers, fruit, birds and insects, and date from various points of her career.

Fig. 36
Jacobus Houbraken after Georg Gsell
A portrait of Maria Sibylla Merian, 1717

Maria Sibylla (fig. 36) grew up in Frankfurt, the daughter of Matthäus Merian the Elder (1593–1650), a successful engraver, publisher and topographical artist, and his second wife. After his death his widow married the Dutch flower-painter and teacher Jacob Marrell (1614–81). Surrounded from an early age by an environment of art and nature study, Merian became fascinated by entomology, recording in her journal years later that she began her investigation of flies, spiders and caterpillars in 1660, when she was thirteen.

In 1665 Merian married one of her stepfather's pupils, Johann Graff, and they moved to her husband's native city of Nuremberg in 1670. Five years later Merian published her first book, *Florum fasciculus primus* (A first bunch of flowers), which she followed with two further parts in 1677 and 1680. These were published in a combined edition in 1680 with the German title *Neues Blumenbuch* (A new book of flowers). The *Blumenbuch* was essentially a pattern-book designed to serve as a model for embroidery; some of the flowers were depicted growing in the earth, while others were arranged as garlands and posies.

Merian's first scientific work, and an important precursor to the *Metamorphosis*, was her *Raupenbuch*, or more fully *Der Raupen wunderbare Verwandelung und sonderbare Blumennahrung* (The wondrous transformation of caterpillars and their remarkable diet of flowers), on which she had worked concurrently with the *Neues Blumenbuch*, and which was published in two parts in 1679 and 1683, with a third part published posthumously in Amsterdam in 1717. Each part comprised fifty plates showing caterpillars, chrysalises, butterflies and moths in their natural habitat, and represented the results of many years of observation. She described it in her introduction as a study 'in which, by an new invention, the origins, food and metamorphoses of caterpillars, grubs, butterflies, moths, flies and other such creatures … are diligently examined, briefly described and painted from life, engraved in copper and published by the author'.

In 1681 Merian's stepfather died, and thereafter she spent much time at her family's home in Frankfurt. During this period she separated from her husband, and in 1685 she moved, with her widowed mother and her two daughters, Johanna Helena, born in 1668, and Dorothea Maria, born ten years later, to Schloss Waltha near the village of Wieuwerd in West Friesland, in order to join a Labadist community to which her half-brother Caspar already belonged. The castle belonged to the Van Sommelsdijk family, the head of which, Cornelis van Aerssen van Sommelsdijk, was the governor of Surinam. He was a follower of the teachings of Jean de Labadie, a French ex-Jesuit Protestant reformer who preached a return to primitive Christianity. The Labadist community had missionary stations on the north coast of South America, and ran a plantation in Surinam. For the next few years at Schloss Waltha, Merian continued to observe insects, remarking in her preface to the *Metamorphosis* that in Friesland she was able to study 'what is found specifically in heath and moorland'.

In addition to the connection with Surinam through the Labadists, the other great impetus for Merian's expedition came in 1691, when on leaving Schloss Waltha after the death of her mother she moved with her daughters to Amsterdam. Thanks to the endeavours of the Dutch East India Company, Amsterdam was at that time the centre of world trade and a site of numerous scholarly collections, many deriving from Dutch colonies. Once there, Merian gained access to the cabinets of some of the leading natural scientists of the city, as she records in her introduction to the *Metamorphosis*: 'In Holland I saw with wonderment the beautiful creatures brought back from the East and West Indies'. Her drawing of the ornate lory (plate 68), which is found in Indonesia, was

Fig. 37
Andries van Buysen, *A cabinet of natural curiosities*
Many of the cases on the tables and on the walls to the right contain entomological specimens.
Frontispiece to Levinus Vincent, *Het Wondertooneel der Natuure*, 1706

evidently made from a specimen she saw in Amsterdam. Of the many collections Merian claimed to have seen, she first listed that of Dr Nicolaas Witsen, one of the most influential men of the city. Burgomaster of Amsterdam and board member of the Dutch East India Company, he possessed an extremely diverse collection, including antiquities, naturalia and ethnographic objects as well as drawings of plants and animals. The second collector she mentioned was Frederick Ruysch, who in 1685 had been appointed professor of anatomy in Amsterdam, and who possessed a large collection of anatomical specimens which he displayed in five rooms of his house. Merian also referred to Levinus Vincent, whose collection, which was likewise open to the public, incorporated numerous snakes and lizards preserved in jars, stuffed birds, shells, crustacea, corals, beetles and a great number of butterflies (fig. 37). These were among the greatest natural history collections of the era.

Although Merian was enthralled by the entomological and other specimens which she saw in collectors' cabinets, her real interest was, as she recalled in her introduction to the *Metamorphosis*, in 'their origins and subsequent development'. She recognised that only fieldwork would provide her with the opportunity to study live specimens and 'to carry out more precise investigations'. So in 1699, at the age of fifty-two, Merian set sail for Surinam with her younger daughter Dorothea in order to study at first hand the country's indigenous butterflies and moths.

Surinam, also known as Dutch Guiana (fig. 38), had come under the control of the Dutch in 1667. Prospective colonists, whose chief concern would be running sugar plantations in the country, were offered incentives, including tax concessions and powers in both civil and penal law; they were also given military protection. First-hand reports of the colony from members of the Labadist

Fig. 38
A map of the coast of Surinam, c.1710

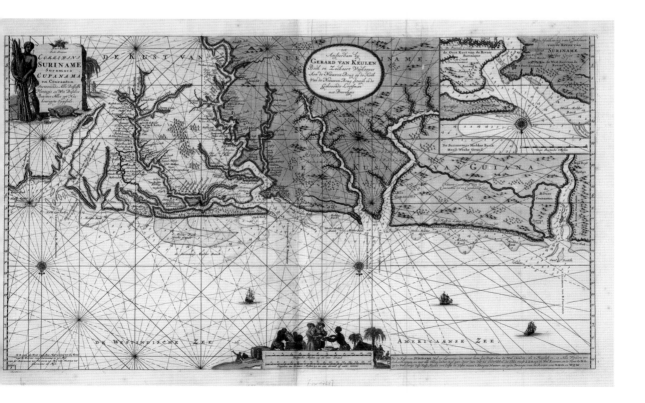

community doubtless gave Merian vital practical information in advance of her visit, yet her courage in making an expedition to Surinam was nonetheless considerable. The colony was relatively newly established and the land largely uncultivated. Moreover, she went on her own initiative, without official patronage or subscribers to an eventual publication of her findings. The conditions which Merian found there were indeed harsh; in a letter written after her return, in 1702, she remarked, '… the heat in this country is overwhelming and one can only work with tremendous difficulty. It nearly cost me my life, which is why I could not stay there any longer. Everyone is amazed that I survived at all. For most people die there of the heat ….'

Our knowledge of Merian's experiences in Surinam is largely provided by her introduction to the *Metamorphosis* and by the textual descriptions which accompany each plate. These remarkably candid and pithy commentaries describe in detail her life in Surinam and her expeditions in search of insects. She repeatedly describes looking for insects and caterpillars in the wild, transporting them back to her garden, feeding them on leaves, drawing them, waiting for them to form cocoons and observing their transformations. She also made forays into the 'wilderness' – the interior of the country – in search of butterflies and plants. Often, however, she found the jungle virtually impenetrable, remarking, 'Yet (in my opinion) one could find a great many other things in the forest if it were passable; but it is so densely overgrown with thistles and thorn bushes that I had to send my slaves ahead with axe in hand to hack an opening for me to proceed even to a certain extent, which nevertheless was very difficult'. Merian also frequently expressed frustration with the ignorance of the Dutch settlers about the country. She was among the few Europeans living in Surinam not involved in sugar cultivation, and she remarks that the colonists 'jeer at me that I am looking for other things than sugar in the country'. She listed a number of crops, including cherries, vanilla, figs and grapes, which she believed could profitably be cultivated if, as she put it, 'the country was inhabited by a more industrious and less selfish population'. Merian's isolation from other natural historians made her task even more difficult; at one point she remarked laconically: 'This plant grew in my garden in Surinam without anyone being able to tell me its name or properties'.

Plate 54. Pineapple (*Ananas comosus*) with Australian cockroaches (*Periplaneta australasiae*) and German cockroaches (*Blatella germanica*), *c*.1701–5

Cassava is also known as manioc and is native to South America. Its roots are still a staple food over large parts of that continent, and it has also been introduced to many other parts of the tropics, notably West Africa. As a food it has the disadvantage, as Merian noted herself, of having an extremely poisonous juice, containing cyanide, which has to be got rid of before cooking.

The moth must have been drawn from a dead specimen, since its proboscis is partially uncurled in a very unnatural way.

The sizes of the different creatures represented here are obviously not relatively correct. A cassava leaf can be nearly 30cm (12 inches) across and the caterpillars even of a sphinx moth do not grow to such a size. Tree-boas on the other hand when adult are six feet or so long. It could be argued that Merian has drawn a newly born youngster but even these are twelve inches long.   D. A.

Plate 55. Roots of the cassava (*Manihot esculenta*) with moth of rustic sphinx (*Manduca rustica*), caterpillar and chrysalis of tetrio sphinx (*Pseudosphinx tetrio*) and garden tree-boa (*Corallus enhydris*), *c.*1701–5

Plate 56. Branch of swamp immortelle (*Erythrina fusca*) with giant silk moths and chrysalises (*Arsenura armida*), *c*.1701–5

Nonetheless, despite the abundant difficulties presented by the climate, lack of access to the interior of the country and a paucity of reliable information, Merian made numerous detailed and exquisitely beautiful watercolour drawings of the insects she found, with accompanying notes, on the small vellum sheets of her *Studienbuch* (now in the Academy of Sciences at St Petersburg; fig. 39). Her delight and fascination with the creatures she found in Surinam is apparent: 'One day I wandered far out into the wilderness …. I took this caterpillar home with me and it rapidly changed into a pale wood-coloured chrysalis, like the one here lying on the twig; two weeks later, towards the end of January 1700, this most beautiful butterfly emerged, looking like polished silver overlaid with the loveliest ultramarine, green and purple, and indescribably beautiful; its beauty cannot possibly be rendered with the paint-brush.'

Merian returned to Amsterdam in 1701 with many preserved specimens in addition to her notes and studies. Her collection included a crocodile, a large variety of snakes, and '20 round boxes with all kinds of butterflies, beetles, hummingbirds, [and] glow-worms'. In a letter of October 1702 to her friend the Nuremberg doctor Johann Georg Volckamer, she explained her method: 'When I was in that country I painted and described the larvae and caterpillars as well as their kind of food and habits; but everything I did not need to paint [there] I brought with me, such as the butterflies and beetles and everything which I could steep in brandy and everything which I could press I am now painting the same way as I did when I was in Germany, but everything on vellum in large format with the plants and creatures life size'.

Using the notes, sketches and dried and preserved specimens which she had brought back from Surinam, Merian now began to design the sixty plates for her great

Fig. 39
Maria Sibylla Merian
*Studies of the metamorphosis
of the giant silk moth, c.1699–1701*

work, the *Metamorphosis insectorum Surinamensium*. Building on the individual studies in her *Studienbuch*, she arranged the caterpillar, chrysalis and moth or butterfly together on the same sheet in a life-cycle tableau. As in her *Raupenbuch*, Merian placed the insects on what she considered to be their food plants. Great emphasis is placed on the brilliance and strange forms of these creatures: Merian remarked to Volckamer that she was painting 'many amazing rare things which have never been seen before'.

The sixty versions of the *Metamorphosis* compositions at Windsor are not, strictly speaking, original designs for the plates of the book. They are among several sets of compositions presumably made during the few years leading up to the publication of the book; another set, purchased by Sir Hans Sloane, is in the British Museum, and a further partial set is in the Academy of Sciences in St Petersburg. On the Windsor sheets, watercolour is applied over faint etched outlines, counterproofed onto the vellum sheets from a freshly printed impression on paper. The etched areas – presumably the etching was carried out by Merian herself

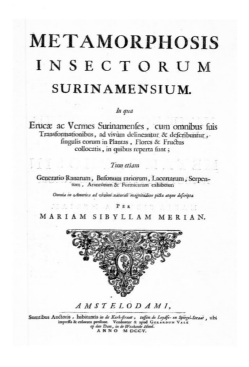

Fig. 40
Title page of the *Metamorphosis insectorum Surinamensium*, 1705

– are principally confined to the insects, which are naturally the most finely detailed areas; leaves and flowers are mostly painted freehand. Merian probably made these compositions as a *de luxe* edition of *Metamorphosis* watercolours.

text continues on page 174

Plate 57. Branch of banana tree (*Musa paradisiaca*) with caterpillar and moth (*Automeris liberia*), *c*.1701–5

OVERLEAF:

Plate 58. Branch of guava tree (*Psidium guineense*) with leafcutter ants (*Atta cephalotes*), army ants (*Eciton* sp.), pink-toed tarantulas (*Avicularia avicularia*), huntsman spiders (*Heteropoda venatoria*) and ruby topaz hummingbird (*Chrysolampis mosquitus*), *c*.1701–5

Two different kinds of ant appear in this drawing. Those in the top half of the picture seem certainly to be leafcutter ants. Most of the leaves, however, are rather improbably nibbled. The ants cannot eat sections from the middle of the leaf leaving the margin entire as Merian shows. Instead, the ants start at the outer margin and work inwards, scissoring the leaf into roughly semi-circular sections. Merian in her text says that the worker ants do not consume the leaf sections themselves – which is correct – but take them back to the underground nest for their young – which is also true but only partially. The leaf segments are taken to underground chambers and there chewed by other workers to produce a mulch from which the ants cultivate a fungus. It is this that is then fed to the young larvae.

In the lower left quarter of the picture Merian figures a different kind of ant. Some of these are winged adults but one is a soldier with greatly enlarged mandibles, more elongated than those of the leafcutters above. Furthermore, this second kind of ant is shown attacking not leaves but another insect. They must therefore be army ants. They do not live in huge underground nests like leafcutters but occupy temporary bivouacs in between their long marches across the forest floor in search of prey. Merian in her text conflates the two kinds. That perhaps is because both kinds can be found running in long columns along established pathways on the forest floor.

Once again, Merian seems little concerned about relative sizes. The hummingbird's nest is represented as being scarcely bigger in diameter than the ants close by. The hummingbird that has been caught by the spider (bottom right) seems a little fanciful. I do not think that there is, in reality, a species with an orange cap and a yellow breast. D. A.

The lowest insect in the picture belongs to a group known as flag-legged bugs. The function of the brightly coloured flanges on the hind pair of legs is not entirely certain, and may, in fact, vary from one species to another. Some apparently wave their legs to scare off possible predators. Others seem to deploy them as deflection devices that induce an attacking bird to peck at a leg rather than the more vulnerable head or abdomen. Merian herself remarked that these legs drop off at a touch, suggesting that this last is in fact the function of the flanges in the species shown here.

D. A.

Plate 59. Passion flower plant (*Passiflora laurifolia*), and flag-legged bug (*Anisoscelis foliacea*) c.1701–5

OVERLEAF:

Plate 60. Vine branch and black grapes (*Vitis vinifera*) with moth, caterpillar and chrysalis of gaudy sphinx (*Eumorpha labruscae*), c.1701–5

Plate 61. Sweet potato plant (*Ipomoea batatus*) and parrot flower (*Heliconia psittacorum*), c.1701–5

The spectacular insects that dominate this drawing are known as lantern flies because early accounts of them declare that the huge proboscis gives off a bright light in the dark. Merian says that it is as bright as a candle and strong enough to read a paper by. She even gives a vivid and seemingly first-hand account of how one night she opened a box in which she had put a large quantity of these flies and a fiery flame emerged. The scientific name of these insects, Fulgora, *stems from the same accounts, for it is derived from Latin* fulgor, *which means 'flash of lightning'. European scientists who bestowed the names on these insects either were working from dead specimens or must have assumed that, if their specimens were alive, they must be ailing, for none can have produced these spectacular lights. No one since has ever seen a light coming from the insects.*

On the other hand, it has to be said that no one has been able to give a convincing explanation of the function of the huge hollow elongations of the fulgorid's head. The fact that in some species the lateral markings look very like the head of a miniature crocodile, complete with a row of teeth, only deepens the mystery. D. A.

Plate 62. Branch from a double-blossomed pomegranate tree (*Punica granatum*) with lantern flies (*Fulgora laternaria*) and cicada (*Fidicina mannifera*), *c*.1701–5

The water hyacinth has air-filled swellings in its stems so that it floats. Perhaps because of the beauty of its lovely lavender-coloured flowers, it was exported from South America to other tropical regions and flourished so greatly in its new locations that it has formed immense floating rafts that cut out all light from the water beneath. It can therefore create problems of industrial proportions on, for example, African lakes such as Kariba.

The giant water-bug is one of the biggest insects in the world, some species growing to 11 cm (over 4 inches) in length. The one shown here grappling with

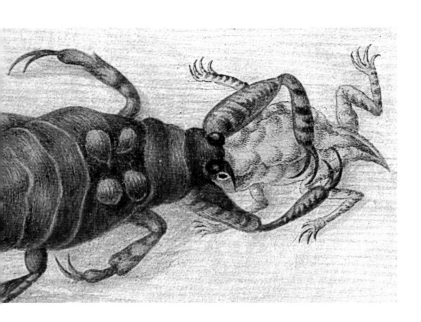

a frog must be a male, for it carries four small spheres on its back. These are eggs which the female sticks to her struggling mate with a water-resistant glue. She produces so many eggs in a season that she needs the services of several males. If her partner has already carried a load, the female will carefully remove the empty shells before fixing on another clutch.   D. A.

Plate 63. Water hyacinth (*Eichhornia crassipes*), marbled or veined tree-frogs with tadpoles and frog-spawn (*Phrynohyas venulosa*) and giant water-bugs (*Lethocerus grandis*), *c.*1701–5

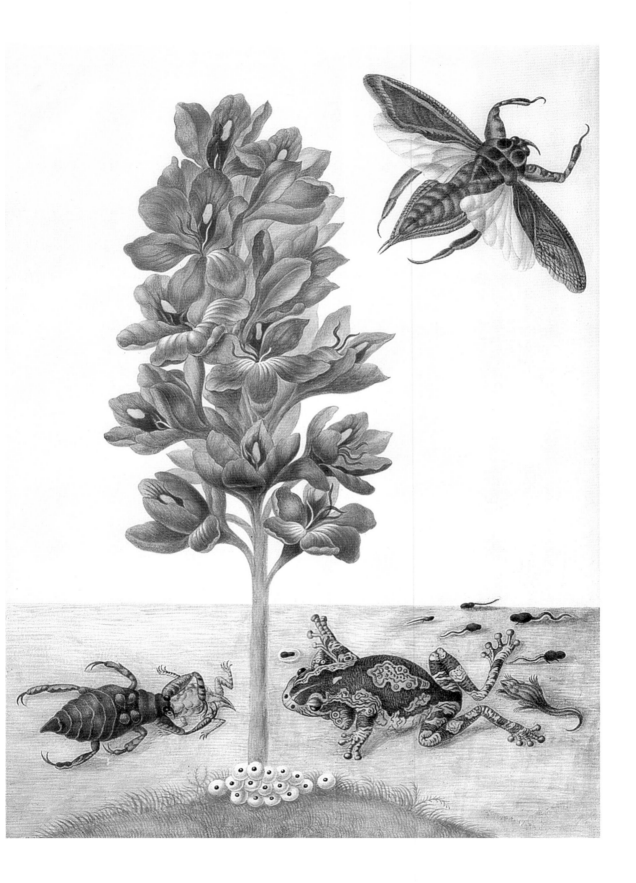

This species of toad has an extremely flattened body. Its most extraordinary characteristic, however, is its method of breeding. Mating begins when the male clasps his forelegs around the female's body just in front of her hind legs. As she extrudes her eggs the male fertilises them and manoeuvres them forward on to her back. There they stick. After she has produced her entire clutch of about sixty, the male disengages and retreats. The skin on the female's back then swells to surround and eventually enclose each egg. Each develops first into a tadpole and then into a tiny froglet which eventually breaks through the mother's skin that enclosed its tiny nursery and swims away. Merian was the first European to describe and publish this strange and unique process.

It is not easy to understand why Merian should have shown this toad with sea shells and a seashore-growing plant since the toad is a freshwater species, found on the Amazon and the Orinoco. D. A.

Plate 64. Sea purslane (*Sesuvium portulacastrum*) and Surinam toad (*Pipa pipa*), c.1701–5

Plate 65. Red-billed toucan (*Ramphastos tucanus*), *c.*1705–10

Plate 66. Aesculapian false coral snake (*Erythrolamprus aesculapii*), banded cat-eyed snake (*Leptodeira annulata*), frog (*Leptodactylus* sp.) and tree frog (*Phyllomedusa tomopterna*), *c.*1705–10

Plate 67. Frog (*Leptodactylus* sp.) with frog-spawn and tadpoles in various stages of development and marsh marigold (*Caltha palustris*), *c*.1705–10

Plate 68. Ornate lory (*Trichoglossus ornatus*) on branch of peach tree (*Prunus persica*), c.1691–9

OVERLEAF:

Plate 69. Common or spectacled caiman (*Caiman crocodilus*) and South American false coral snake (*Anilius scytale*), c.1705–10

Plate 70. Golden tegu lizard
(*Tupinambis nigropunctatus*), *c*.1705–10

Plate 71. Still life with flowers tied at the stems, *c.*1705–10

Plate 72. Still life with fruit and blue–backed manakin (*Chiroxiphia pareola*), *c.*1705–10

The *Metamorphosis insectorum Surinamensium* was published in 1705 (fig. 40). It was aimed at a broad market; dedicated in the preface both to 'lovers of art' and to 'lovers of insects', it was conceived as a beautiful and desirable book with an appeal to the non-specialist as well as to the naturalist community. Four further editions of the *Metamorphosis* were published after Merian's death. These included twelve additional plates depicting not only more butterflies but also Surinamese reptiles, amphibians and marsupials, the most spectacular of which was a magnificent spectacled caiman fighting with a coral snake (plate 69).

As an entomologist, Merian built on the great developments in the study of insects which had taken place during the seventeenth century. Entomology had been established as a science in its own right around a hundred years previously, with Ulisse Aldrovandi's publication in 1602 of the pioneering work *De animalibus insectis*, the first book to systematise insect taxonomy. In her preface to the *Metamorphosis* Merian refers to the work of four renowned seventeenth-century entomologists, including Johannes Goedaert and Jan Swammerdam. Goedaert's three-volume work *Metamorphosis naturalis* (1662, 1665 and 1669) had transformed knowledge of the metamorphosis process by making, for the first time, an empirical study of the stages of development from egg to imago. In his *Historia insectorum generalis*, published in Dutch in 1699 as *Algemeene verhandeling van de bloedeloose dierkens*, Swammerdam had sought to disprove the ancient theory that insects were spontaneously generated, arguing for the continuation of the organism through the process of metamorphosis. Thus Merian was by no means the first to understand the process of insect metamorphosis, or to employ an empirical approach to entomology. However, her great innovation – and what made her book appeal to the 'lover of art' as well as the specialist – was to show insects in a naturalistic context, with the plants upon which they fed, against the prevailing convention of representing the caterpillar, chrysalis and moth or butterfly in a diagrammatic way, in neat rows and isolated from their natural habitat.

Merian's compositions for the *Metamorphosis* constitute a significant departure from her previous work in their greater boldness, both of colour and form. Where previously her compositions had been relatively restrained, the exotic plants and flowers of this book are shown in vibrant displays of twisting, spiralling and highly coloured forms, often exceeding the edges of the sheet. Another striking development is the macabre, even nightmarish quality of some of her compositions and the introduction of a dramatic narrative, particularly evident in plate 58, which shows a pink-toed tarantula poised to eat a hummingbird, and in plate 63, in which a giant water-bug devours a frog. The *vanitas* theme in Dutch still-life painting, absent from the *Raupenbuch* and Merian's previous work, emerges in these tableaux. Arguably, this change of emphasis had its origins not directly in her knowledge of contemporary still-life paintings, with which, through her stepfather, she must have been familiar from a young age, but in the arrangements of some of the cabinets of curiosities which she had seen in Amsterdam prior to her expedition. Frederick Ruysch's collection in particular had a strong emphasis on the *vanitas* theme, and included spectacular groups of preserved human organs flanked by skeletons of small children so as to form elaborate allegories of death, of which he had engravings made (fig. 41).

Fig. 41
Cornelis Huyberts
Frontispiece of Frederik Ruysch,
Thesaurus anatomicus, 1701–06

The ninety-five watercolours by Merian now in the Royal Library were purchased in 1755 by George III, then Prince of Wales, from the posthumous sale of the collection of Dr Richard Mead (1673–1754), a renowned bibliophile and collector. The two large leather bindings in which the drawings have been housed for many years probably date from the period of Dr Mead's ownership; the original bindings, which might have borne some evidence of the volumes' earlier history, no longer survive. It is not clear whether Mead had commissioned these works directly from Merian, or whether he acquired them through agents working for him on the Continent.

RK CATESBY

SUSAN OWENS

'A GENIUS FOR
NATURAL HISTORY'

MARK CATESBY (1682–1749) published his life's work, *The Natural History of Carolina, Florida and the Bahama Islands* in parts between 1729 and 1747. This magnificent two-volume study, comprising full-page illustrations with accompanying textual descriptions, was the first on its subject. The 263 original watercolours by Catesby in the Royal Collection – the main surviving group of Catesby's work – were preparatory studies for the plates of this publication. Made during the artist's two extended visits to the east coast of North America, first to Virginia, then to Carolina, these watercolours depict numerous different species of plants, birds, fish and snakes. Catesby intended nothing less than a comprehensive survey of the flora and fauna native to the eastern seaboard of North America (fig. 42).

The burgeoning garden culture of seventeenth-century Britain, fuelled by the rapid introduction of plant species from the Near East at the end of the sixteenth century, had created an avid curiosity about the botanical products of the New World. Of particular allure was Virginia, where in 1607 Jamestown, the first permanent English settlement in North America, had been founded. The collector and gardener John Tradescant the Younger (see p. 118) made three expeditions to Virginia, and he brought back a number of species including the American plane tree, the tulip tree and the Virginia creeper. One of the first botanists to study the indigenous species of this region was the Revd John Banister (1652–1692), who was sent to Virginia in 1678 by Henry Compton, Bishop of London (see p. 129). Banister collected and drew many plants, sending a great number back to Compton's garden at Fulham Palace, including the sweet bay, the first magnolia to be introduced to Britain. His discoveries were published by the renowned naturalist John Ray (1627–1705) in the second volume of his seminal work on British flora, *Historia plantarum* (1688), but without illustrations. The priest Charles Plumier (1666–1706) published four significant works on the plants of America at the end of the seventeenth and in the early years of the eighteenth century. However, although pioneering botanists before him had initiated the study of the plants which were to be found in North America, Catesby was the first naturalist to extend his attention to the animals and birds which coexisted with them, and to indicate their mutual dependence.

Catesby was born in the town of Sudbury, in Suffolk. Although a certain amount of biographical information is provided by Catesby himself in the preface to the *Natural History*, little is known about his early life and education. His youthful interest in natural history, what he calls his 'early Inclination … to search after Plants, and other Productions in Nature', was encouraged by his uncle Nicholas Jekyll, an associate of John Ray. As a friend of Catesby noted later, Jekyll inspired in him 'a genius for natural history'.

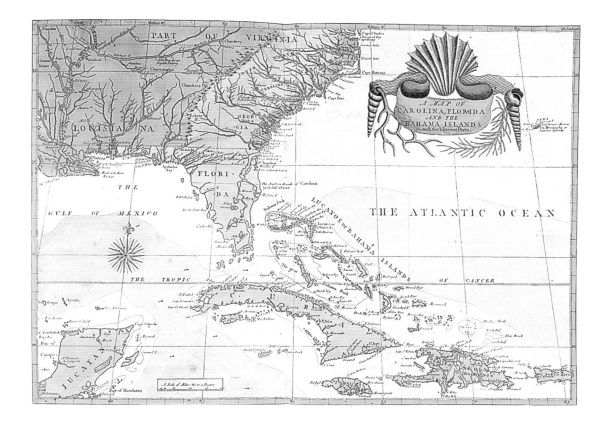

Fig. 42

A map of part of Virginia, Carolina, Florida and the Bahama Islands bound into Catesby's *Natural History*

At first Catesby's botanical studies were, as he notes in the preface to the *Natural History*, 'much suppressed by my residing too remote from *London*, the Center of all Science'. However, a great opportunity to continue his studies further afield and to observe plants and animals which were 'Strangers to *England'* presented itself through a family connection. His brother-in-law, Dr William Cocke, had established a medical practice in Williamsburg, the capital of the Virginia colony, and in 1712 Catesby accompanied his sister Elizabeth when she travelled out to join her husband. Cocke, who was also a politician, was well connected in Virginia, and he was able to introduce Catesby to wealthy landowning friends, many of whom, like their English counterparts, were interested in horticulture and were keen to assist Catesby in his research. For the following seven years Catesby explored the Tidewater plantations of Virginia and made expeditions up the James river towards the Appalachian mountains. The main purpose of these excursions was to collect botanical specimens and seeds for Samuel Dale, a colleague of John Ray, and for Thomas Fairchild, the owner of an experimental nursery at Hoxton and the author of the *City Gardener* (1722).

This is an unusual plate. Catesby not only shows his subject in flight instead of standing in profile but he also includes a landscape background. Furthermore the illustration depicts a drama involving another species of bird. An osprey is shown, a little more distantly, at top right. It is an expert at seizing fish from the surface of the water but it has dropped one, perhaps after being harried by the bald eagle which is now seizing the fish in mid-air.   D. A.

Plate 73. Bald eagle (*Haliaeetus leucocephalus*), *c.*1722–6

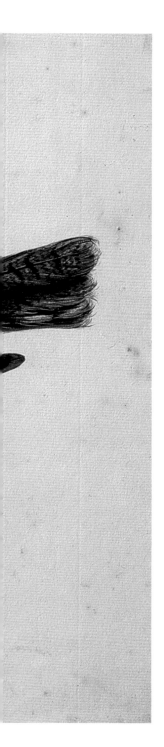

Nightjars have a worldwide distribution and Catesby would have had no difficulty in naming this species since it is very similar to the European species he must have known well. The bird also has the name goatsucker (their scientific name Caprimulgus *derives from Latin words meaning the same thing) because of a myth that the birds take milk from goats. It is true that they are often seen fluttering through the darkness around pens where goats or other domestic livestock are kept for the night, but in fact they come to collect insects that may have been attracted by the animals' droppings.*

Nightjars usually collect their insects in flight, with the aid of the sensory bristles that fringe the beak. They may occasionally take prey from the ground but the mole cricket drawn by Catesby is an unlikely food for them, for it spends most of its time in a tunnel which it digs with its powerful forelegs. There it feeds on worms, insect larvae and occasionally a few plant roots. It only emerges from underground for any length of time to seek a mate. D. A.

Plate 74. Nightjar (*Caprimulgus carolinensis*) and mole cricket (*Gryllotalpa gryllotalpa*), c.1722–6

*The ivory-billed woodpecker, the largest species in the woodpecker family, is a
spectacular bird the size of a raven. Catesby could have seen it in most parts of Florida
as well as the Bahamas. It was also to be found at that time in Cuba. By the 1950s
it had disappeared from most of its original habitat and by the 1970s it was considered
almost certainly to be extinct. Recently, however, there has been a sighting at a secret
location, which gives the hope that the species may survive after all.*   D. A.

Plate 75. Ivory-billed woodpecker (*Campephilus principalis*) and willow oak (*Quercus phellos*), *c*.1722–6

Quercus angustis
Augusto Salicis
Aler. Marilandica folia longo
Ri. Hist: — Willow Oak.

Prompted by the first volume of Sir Hans Sloane's *A Voyage to the Islands Madera, Barbados, Nieves, St Christophers and Jamaica*, published in 1707, in 1714 Catesby also travelled to Jamaica in order to study the flora and fauna of the West Indies.

Catesby later regretted what he considered to be his unmethodical approach during this first visit to America, writing in the *Natural History*: 'I thought then so little of prosecuting a Design of the Nature of this Work, that in the Seven Years I resided in that Country, (I am ashamed to own it) I chiefly gratified my Inclination in observing and admiring the various Productions of those Countries ... only sending from thence some dried Specimens of Plants and some of the most Specious of them in Tubs of Earth, at the Request of some curious Friends ...'. However, without doubt these years in the field were of enormous value when he returned to America several years later.

Back in England in 1719, Catesby showed his friend Samuel Dale a group of drawings which he had made while in Virginia. Dale arranged for Catesby to meet the celebrated botanist William Sherard, writing, 'Mr. Catesby is come from Virginia ... he intends againe to return, and will take an opportuniity to waite upon you with some paintings of Birds &c. which he hath drawn. Its [a] pitty some incouragement can't be found for him, he may be very usefull for the perfecting of Natural History'. Sherard was greatly impressed by Catesby's drawings, remarking, 'He designs and paints in water colours to perfection'. As it happened, Sherard was already discussing the possibility of sending a naturalist to America, and now organised a group of subscribers to send Catesby back across the Atlantic – this time to Carolina – on a salary of £20 a year. Although the Royal Society gave Catesby no financial backing it gave him valuable support, enabling him to attract wealthy subscribers. Sir Hans Sloane, founder of the British Museum, was one of the most important patrons for this second visit; another was the physician and collector Dr Richard Mead who also owned the large group of drawings by Maria Sibylla Merian which are now in the Royal Collection – see pp. 138–75).

In 1722 Catesby set sail for Charleston. At this time South Carolina was still frontier country. He notes in his preface to the *Natural History* that the natural resources of the colony had been little explored beyond those of commercial interest such as rice, pitch and tar. However, he describes the country as 'inferior to none in Fertility, and abounding in Variety of the Blessings of Nature'. On his arrival he was welcomed by General Nicholson, governor of the province, who, like Catesby's brother-in-law on his previous visit to America, arranged for his introduction to influential members of the colony.

Plate 76. Passenger pigeon (*Ectopistes migratoria*) and turkey oak (*Quercus laevis*), *c*.1722–6

In Catesby's time the passenger pigeon was one of the commonest birds in North America. Some authorities have even suggested that it was the most numerous bird that ever existed. Immense flocks, one estimated at containing two thousand million individuals, flew over the grasslands of central North America, darkening the skies and taking three days to fly past. As people settled on the prairies, the flocks began to decline in size. Suddenly the bird became very rare and the last wild individual was sighted in 1889. The last survivor of the species, a lonely captive female named Martha, died in the Cincinnati Zoo in 1914.

Catesby's indifference to relative sizes allowed him to place his bird on an oak leaf that is almost as long as the bird itself. In fact the passenger pigeon was no smaller than many other members of the pigeon family.   D. A.

Plate 77. Red-legged thrush (*Turdus plumbeus*) and gumbo limbo tree (*Bursera simaruba*), *c*.1722–6

Plate 78. Yellow-throated warbler (*Dendroica dominica*), pine warbler (*Dendroica pinus*) and red maple (*Acer rubrum*), *c*.1722–6

Grey Titmous with a Yellow Throat   61   62   Yellow Throat
Parus Americanus cinereus .   Pine Creeper

Pine creeper   p. 62
Parus Americanus Lutescens   Acer Virginianum folio majore, subtus argenteo, supra
                                    viridi Splendente . . . Pluk: Alma.

During this expedition Catesby's approach was far more systematic than it had been on the previous occasion, and he arranged his travels to various part of the country in order that he should see them during different seasons, as he described in a letter to Sherard:

My method is never to be twice at [the same] place in the same season for if in the sp[ring] I am in the low Country [in the Sum]mer I am [at] the hea[ds] of rivers the next Summer in the low countrys, so alternating that in 2 Years [I visit] the two different parts of the Country.

Catesby describes his exploration of the inhabited coastal plain of Carolina, and how he spent the first year 'searching after, collecting and describing the Animals and Plants'. He then travelled to the uninhabited areas of the country around Fort Moore, a small fortress on the banks of the Savanna. He was delighted to find there 'abundance of Things not to be seen in the Lower Parts of the Country', and was prompted by his discoveries to 'take several Journeys with the *Indians* higher up the Rivers, towards the Mountains'. He records that these expeditions 'afforded not only a Succession of new vegetable Appearances, but [the] most delightful Prospects imaginable, besides the Diversion of Hunting Bufello's, Bears, Panthers, and other wild Beasts'. He goes on to describe how 'In these excursions I employ'd an *Indian* to carry my Box, in which, besides Paper and Materials for Painting, I put dry'd Specimens of Plants, Seeds, &c – as I gather'd them', adding 'to the Hospitality and Assistance of these Friendly *Indians*, I am much indebted'.

Catesby's greatest interest was in botanical specimens, particularly trees and shrubs, not only for their intrinsic interest but also for their 'several Mechanical and other Uses, as in Building, Joynery, Agriculture, and others used for Food and Medicine'. Of principal interest were trees which might be imported to England and successfully cultivated there. He also made an intensive study of birds, explaining his preference thus: 'There being a greater Variety of the feather'd Kind than of any other Animals … and excelling in the Beauty of their Colours, besides having oftenist relation to the Plants on which they feed and frequent'. In 1725 Catesby made a trip to the Bahamas, where he concentrated on fish, explaining that he had deferred their study until this visit. He was not disappointed: 'tho' I had been often told they were very remarkable, yet I was surprised to find how lavishly Nature had adorn'd them with Marks and Colours most admirable'.

According to Catesby's description of his working practices, he made watercolour studies in the field: 'In designing the Plants, I always did them while fresh and just gather'd: And the Animals, particularly the Birds, I painted them while alive (except a very few) and gave them their Gestures peculiar to every kind of Bird'. The colours of fish fade rapidly when removed from the water, and Catesby records that he had 'a succession of them procur'd while the former lost their Colours'.

Shortly after his return to England in 1726 Catesby began the project which would occupy the rest of his life – publishing his drawings and observations in the form of a natural history book. In order to support himself in this venture Catesby worked in the Hoxton nursery of Thomas Fairchild, the nurseryman to whom he had sent botanical specimens from America, and later in the nursery of Christopher Gray in Fulham.

When preparing the plates for the *Natural History* Catesby found his preparatory drawings invaluable. He wrote: 'Should any of my original Paintings have been lost, they would have been irretrievable to me, without making another voyage to *America*, since a perpetual inspection of them was so necessary towards the exhibition of truth and accuracy in my descriptions'. As careful studies of specimens found in the field, their purpose was clarity and accuracy rather than artistry; essentially they were technical drawings. In his preface to the *Natural History* Catesby states this plainly: 'As I was not bred a Painter I hope some faults in Perspective and other Niceties may be more readily executed, for I humbly conceive Plants, and other Things done in a Flat, tho' exact manner, may serve the Purpose of Natural History, better in some Measure than in a more bold and Painter like Way'.

Futhermore, Catesby was convinced that plates were of far greater value than any textual description could be:

The Illuminating [of] Natural History is so particularly Essential to the perfect understanding of it, that I may aver a clearer Idea may be conceiv'd from the Figures of Animals and Plants in their proper Colours, than from the most exact Description without them: Wherefore I have been less prolix in the Description, judging it unnecessary to tire the Reader with describing every Feather, yet I hope sufficient to distinguish them without Confussion.

text continues on page 210

This is the greater flamingo, the biggest and most widespread of the five species of flamingo which are found in Europe and Africa as well as the Bahamas, where Catesby saw it. His decision to devote a special drawing to its head shows that he was well aware of its unique structure. Internally, the beak is lined with rows of thin plates fringed with hairs. The bird draws water into its mouth and forces it out again by pumping movements of its tongue, in the process filtering out the small organisms on which it feeds. From a food-gathering point of view, flamingos are the avian equivalents of baleen whales.

The tree-like structure that forms the background is a gorgonian, a colonial marine animal sometimes known as a horny coral. It is very unlikely to occur in the shallow lagoons frequented by flamingos. D. A.

Plate 79. Head of the flamingo (*Phoenicopterus ruber*) and gorgonian (*Plexaura flexuosa*), *c.*1725

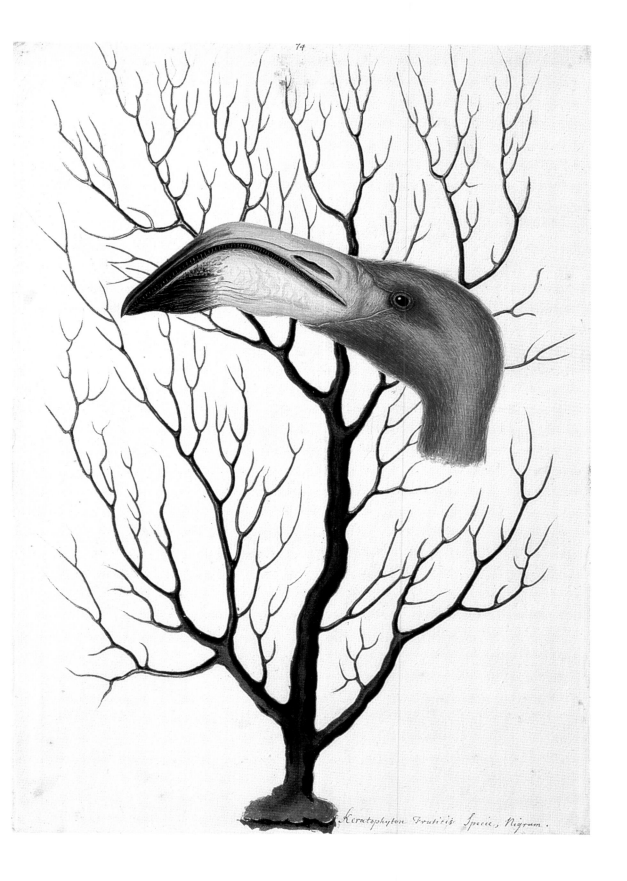

Keratophyton Fruticis Specie, Nigrum.

Catesby writes in the introduction to his book that 'Fish which do not retain their colours when out of their Element, I painted at different times, having a succession of them procur'd while the former lost their Colours'. This drawing is splendid and incontrovertible proof that he did so. D. A.

Plate 80. Great hogfish
(*Lachnolaimus maximus*), *c.*1725

Plate 81. Angel fish
(*Angelichthys ciliaris*), *c.*1725

These are plainly drawn from dead specimens: the small triangular section at the back (which is a much reduced abdomen) is normally held forward, clasped against the main part of the body. The legs too are unnaturally splayed.   D. A.

Plate 82. Sally lightfoot or rock crab (*Grapsus grapsus*) and flame box crab (*Calappa flammea*), *c.*1725

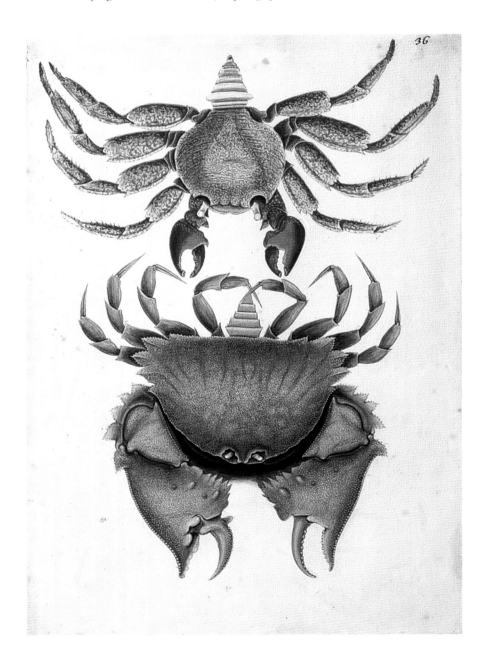

This is the Magnolia grandiflora *that is now planted so widely in Britain. Catesby has drawn the cone-like structure that develops after a flower has shed its petals. As it grows, its seeds fall out but they are suspended by silky threads. Birds are thus able to collect them with ease. They digest the fleshy outer covering of the seed. The centre of the seed, however, remains undamaged and will germinate when the bird eventually voids it.* D. A.

Plate 83. Bull bay magnolia (*Magnolia grandiflora*), *c*.1722–6

Plate 84. Striped skunk
(*Mephitis mephitis*), *c*.1722–6

62

Although this little lizard is commonly called a 'chameleon' it is not closely related to the true chameleons, which are primarily African and Madagascan. It belongs instead to the New World group of iguanas, and is known as an anole. Catesby here has portrayed the relative proportions of animal and plant more or less correctly, for anoles are sufficiently small and agile to clamber through the leaves and stems in the way he shows. The male has a scarlet throat pouch which it can keep concealed or flick forwards in the way Catesby shows, as a territorial display to keep rivals away from its patch in the tree.   D. A.

Plate 85. Jamaican anole (*Anolis garmani*) and sweet gum (*Liquidambar styraciflua*), *c.*1722–6

63

Liquid-Ambari arbor Stiraciflua Aceris folio
fructu tribuloide (sic) pericarpio Orientari ex quam
plurimis apicibus coagmentato Semen recondens
Phutoar: Pluk: Alma: .

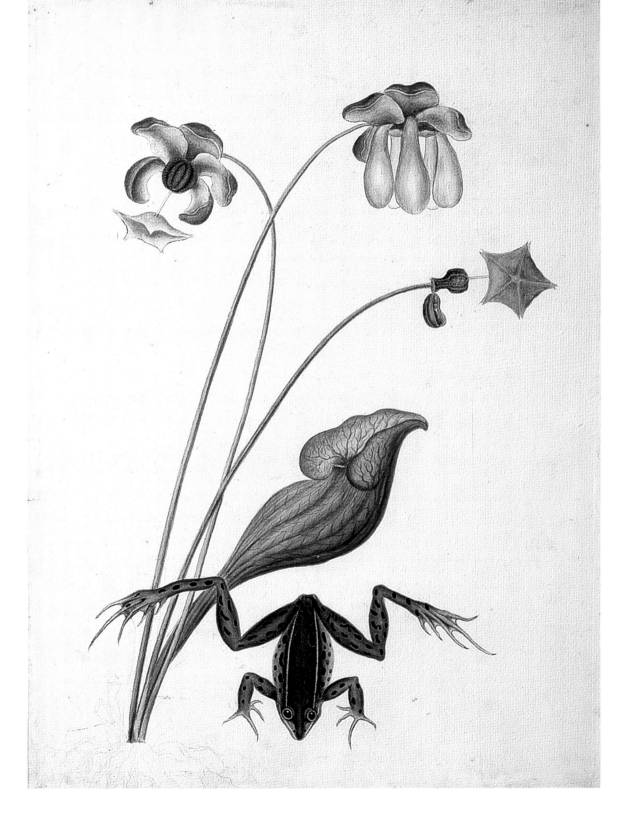

*Sarracenias are carnivorous plants. Insects are attracted by a sweet glistening fluid
that is excreted from the top of the trumpet. But the inner surface of the trumpet is
slippery and insects fall from it to drown in the fluid at the bottom. There their bodies
are digested. Small frogs often squat just inside the top of the trumpet and somehow
manage to keep a foothold as they wait to snap up insects attracted to the plant.
It is doubtful, however, whether this species of frog does so. It looks too big.* D. A.

Plate 86. Water frog (*Rana esculenta*) and purple pitcher plant (*Sarracenia purpurea*), *c.*1722–6

Plate 87. Pitch apple (*Clusia rosea*), *c*.1725

This plant starts life as a seed scraped on to the branch of a tree by a bird or a bat. It sends down long roots to the ground, develops a large crown and eventually strangles and kills the tree on which it sits, continuing life as an independent tree standing sometimes sixty feet (18m) high. The resin from its stem is used to caulk boats.   D. A.

OVERLEAF:

Plate 88. American bison (*Bison bison*) and bristly locust (*Robinia hispida*), *c*.1722–6

Here Catesby's disregard for comparative sizes reaches surreal levels. Why he should have represented North America's biggest animal, a 1,000-kilo bull bison, leaping skittishly through a twig of a Robinia tree is baffling. The animal, of course, in Catesby's time still roamed in vast herds across the North American prairies.   D. A.

Cenchramidea Arbor Saxis adnascens. Obrotundo pingui folio, fructu pomiformi in plurimas capsulas
granula ficulnea stilo columnari hexagono præduro adhærentia continentes diviso, Balsamum fundens.
Pluk: Alma.